"In this comprehensive textbook, readers will discover a wealth of stretching techniques designed to improve flexibility, sport performance, prevent injuries, and promote overall health. Whether you are an athlete looking to enhance performance or simply aiming to increase your range of motion, Professor Behm's textbook is a must-have addition to your fitness library."

Prof. Urs Granacher PhD, University of Freiburg, *Germany*

"Professor Behm shares his profound knowledge and expertise in this book. The body of available evidence is presented using an explicative and practical narrative. An essential read for all those interested in the topic."

Prof. Rodrigo Ramirez-Campillo PhD, Exercise and Rehabilitation Sciences Institute. School of Physical Therapy. Faculty of Rehabilitation Sciences. Universidad Andres Bello. Santiago, *Chile*

"It is the perfect book to tap into the wisdom of stretching as developed by researchers. By reading this book, you will gain wisdom about the past, present, and imagine the future of stretching"

Prof. Masatoshi Nakamura PhD, Faculty of Rehabilitation Sciences, Nishi Kyushu University, Kanzaki, Saga, *Japan*

"With this work, Prof. Behm provides an encyclopedia of flexibility techniques that illustrates the current scientific knowledge on stretching and related training techniques in the most comprehensive manner. The content of this book served as an important baseline for much ongoing research, including my PhD Thesis."

Konstantin Warneke PhD, University of Graz, Graz *Austria*

The Science and Physiology of Flexibility and Stretching

The Science and Physiology of Flexibility and Stretching is the most up-to-date and comprehensive book to cover the underlying physiology and psychology of flexibility enhancing techniques, critically assessing why, when, and how we should stretch, as well as offering a highly illustrated, practical guide to stretching exercises. This fully revised new edition not only updates the present information but adds new chapters on areas that have attracted substantial interest in the past five years such as foam rolling, vibration, global effects of stretching, alternative methods to increase flexibility (e.g., resistance training), and others.

Richly illustrated and including an online resource, *The Science and Physiology of Flexibility and Stretching* provides an important scientific inquiry into stretching as well as other flexibility enhancing techniques (e.g., foam rolling, vibration, resistance training, and others) and an invaluable reference for any strength and conditioning coach or student, personal trainer, sports coach, or exercise scientist.

David Behm is a University Research Professor at Memorial University of Newfoundland. He was a highly competitive athlete, excelling in baseball and ice hockey, and he was drafted into the Canadian Football League (1979). His athletic background led him to seek out an academic career in the areas of applied neuromuscular physiology and sport/exercise science. Dr. Behm has won a number of university, national, and international awards based on his internationally recognized research contributions. At the time of publication, he has published more than 20 book chapters and over 400 peer-reviewed scientific articles, which have been cited more than 35,000 times. He consistently presents his research findings internationally, and his work is often featured in popular fitness and health magazines and online publications.

The Science and Physiology of Flexibility and Stretching

Implications and Applications in Sport Performance and Health

Second edition

David Behm

NEW YORK AND LONDON

Designed cover image: Getty images

Second edition published 2025
by Routledge
605 Third Avenue, New York, NY 10158

and by Routledge
4 Park Square, Milton Park, Abingdon, Oxon OX14 4RN

Routledge is an imprint of the Taylor & Francis Group, an informa business

First edition published by Routledge 2018

Library of Congress Cataloging-in-Publication Data
A catalog record has been requested for this book

ISBN: 978-1-032-72561-1 (hbk)
ISBN: 978-1-032-70907-9 (pbk)
ISBN: 978-1-032-70908-6 (ebk)

DOI: 10.4324/9781032709086

Typeset in Times New Roman
by Taylor & Francis Books

Contents

Figures

Tables

Section I

The Science and Physiology of Stretching

1 My Personal Motivation for Stretching

The year is 1972. A 15-year-old, 5 foot, 10-inch (1.78 meters), 175-lb (79.4 kg) fullback takes his stance three yards behind the quarterback. This grade 11 fullback playing junior high school football is bigger and stronger than most of the offensive linemen blocking for him as well as most of the young adolescent opponents who will try to tackle him. With the ball snapped on the second "hut", the quarterback rotates and hands the ball off to the full-back who is accelerating to an expected opening between the tight end and tackle. As the tight end cross blocks upon the defensive end and the offensive tackle pulls out against the linebacker, a sliver of daylight appears. I lower my shoulder and plunge through that hole. Arms reach out from the partially blocked defensive lineman and linebacker but they are not strong enough to slow me down. A strong side defensive back at about 140 pounds (63.5 kg) moves in for the tackle. However, with the momentum of my greater mass and the velocity attained after an 8-yard sprint, the defensive back is trampled and I cut sharply to the sidelines. After covering about 20 yards, the defensive safety catches me from behind and trips me up.

The next year, I am the starting fullback for the senior high school team. I have grown ¾ of an inch (2 cm) and now weigh 185 lbs (84 kg). We win the regional high school championship. Dave Behm (The Truck) and Dan (Crazy Legs) Murphy make a great one-two punch. My predominant empire is inside between the two ends. Dives and off-tackle plays are my bread and butter. If I get a decent block and get into the defensive backfield, I can use my size advantage, my balance, and my signature move: hit the opponent at full throttle to either knock him down or spin immediately after contact so that the enemy cannot easily grab me and pull me down. Murphy's territory is outside the ends with sweeps and pitches, as he is lighter and much swifter than I and can often outsprint the defence.

In the last year of high school, there is no increase in height but I continue to fill out, expanding to 198 lbs (90 kg). A number of Canadian universities attempt to recruit me and I decide to stay home to play with the University of Ottawa Gee-Gees, who had lost that year in the national semi-final game. However, as the next year approaches, I realize that my chances of getting into a game with that team are slim to none. The starting fullback is Neil

DOI: 10.4324/9781032709086-2

Lumsden, a 235 lb (106 kg) behemoth with decent speed, great balance, strength, and power. He will set Canadian university rushing and touchdown records that will remain untouched for a couple of decades and he will subsequently establish a long career in the professional Canadian Football League (CFL). The back-up is Mike Murphy, another talented fullback at just under 230 lbs (104 kg) who in the following year will lead the nation in rushing yards and also have a firm career in the CFL. I decide to play for the city's junior football team (Ottawa Sooners) and wait for Lumsden to move onto the CFL. Lumsden and Murphy were archetype fullbacks, massive, strong, and powerful. Relatively, I filled that description in high school but my growth pattern started to plateau, such that I was around 205 lb (93 kg) when I joined the university football team in my second year of university. I spent my second year of university primarily blocking for the burly Mike Murphy and moved into the fullback position in my third academic year. I was lucky enough to inherit a strong offensive line and with my slashing, spinning style I could often break a few tackles and make a major gain (Figure 1.1). With my lack of breakaway running speed, I was typically caught from behind by a fleet footed defensive back. For a professional running back, my size was more typical of tailbacks or halfbacks depending on your terminology. These backs typified by legends like Walter Payton had very good to great speed that would allow them to burst into the open field

Figure 1.1 David Behm (32) caught "again" from behind

and outsprint the opposition. Unfortunately, I was built like a tailback but with the speed of a fullback. I needed to get faster if I wanted to continue my career after university.

Sprinting speed is a simple combination of stride rate (frequency) and stride length (7). Stride rate is very difficult to modify as it is generally related to your genetic profile of fast-twitch to slow-twitch muscle fibre composition. Pick the right parents, who hopefully will pass on a higher percentage of fast-twitch fibres and you will be able to move your legs back and forth (stride rate) much quicker than someone with a greater percentage of slow-twitch fibres. Hence, you will have a high stride rate. This is the most important factor in sprinting speed. Unfortunately, it seems that I did not pick the right parents! Thus, I was left with trying to enhance the second sprinting factor: stride length. With stronger more powerful legs, it should be possible to explode off the ground and cover greater distances with each stride. With this in mind, I worked faithfully on my strength, such that I could squat over 500 lbs (225 kg) and bench press around 350 lb (160 kg). As a university student in physical education in the late 1970s, I was taught that by increasing my flexibility I would improve performance (increase stride length, decrease resistance to stride movements) and decrease the chances for injury.

With these pearls of wisdom in mind, I also worked diligently on my stretching so that I could eventually perform a front split. Conventional wisdom of the time indicated that with higher levels of flexibility there would be less resistance to movement and an increased efficiency of movement. Thus, with my improved power and flexibility my stride length should have been tremendous. Like Superman, I should have been leaping over tall buildings in a single bound. However, there must have been some kryptonite in my diet since, while my sprinting speed did improve marginally, nobody ever mistook me for the legendary Walter Payton (or Superman!).

Well if I did not get much faster, then my flexibility should have decreased my chances for injury. In my second year of university football, I took the ball on a draw play (quarterback fakes a pass and then gives the ball to the running back) and burst through a gap in the line. Quickly a linebacker exploded from the left accelerating his helmet into my shoulder. I tried to absorb the hit and I bounced off the hit and continued for another 12 yards till I was, as usual, caught from behind by a defensive back. Upon getting up I noticed my clavicle was apparently redirected towards my back and no longer attached to my scapula. When I returned to the sidelines I was informed that I had subluxated my acromio-clavicular joint. I was out for the season.

With aggressive rehabilitation and off-season training I was ready for my third year of university football. In a game against McGill University on artificial turf, I caught a flare pass and sprinted wide. As I planted my right foot to move up field I was hit low on the left side by a linebacker and simultaneously hit high from the right by a defensive back. My planted foot could not move or slide on the artificial turf and the ligaments were torn resulting in a third-degree ankle sprain. It did not seem that my extensive stretching programme had provided me the protection I sought from musculotendinous or ligamentous injuries nor did it provide me with significantly better athletic performance (improved speed). Did my physical education professors of the 1970s really know what they were talking about? The paradigm of stretching and flexibility has experienced a number of shifts in the past few decades.

It is now about 50 years later (2024) and I am no longer a young athlete but I still try to maintain my fitness (Figure 1.2). Although I am officially a senior citizen (>65 years), I have not lost my competitive nature and still seek to improve my athletic skills in tennis and ice hockey (my arthritic knees make squash more difficult now!) so I can compete with and beat those youngsters (anybody under 50 years old) (Figure 1.3). However, my preparations and warm-ups have been altered by our research findings on flexibility and pre-activity preparations. As we now know that resistance training can provide similar increases in range of motion as static stretching (1), there is no longer a need to stretch before lifting weights. However, I still include static stretching (less than 60 seconds per muscle group: 3, 4, 5, 6) before I play tennis or ice hockey. Periodically if I really feel stiff, I might add foam rolling (2) to my regimen.

Hopefully the second edition of this book will update and help establish the facts about flexibility as they are presently known and burst the myths. It is the

Figure 1.2 Old David Behm trying to maintain fitness and fighting the ravages of ageing (65 years old). Possible positive effects of functional work, resistance training, competitive sport, and maybe stretching?

Figure 1.3 Old David Behm still competing in as many sports as possible. Did stretching contribute to my relative athletic success as a senior citizen?

objective of this book to provide you with the most up to date research on flexibility enhancing exercises and devices as well as explaining the physiological mechanisms underlying different types of flexibility-improving exercises/devices and then provide you with suggestions for appropriate activity programmes to increase your range of motion.

References

1. Alizadeh, S., Daneshjoo, A., Zahiri, A., Anvar, S.H., Goudini, R., Hicks, J.P., Konrad, A., and Behm, D.G.Resistance Training Induces Improvements in Range of Motion: A Systematic Review and Meta-Analysis. *Sports Med* 53: 707–722, 2023.
2. Behm, D.G., Alizadeh, S., Hadjizadeh Anvar, S., Mahmoud, M.M.I., Ramsay, E., Hanlon, C., and Cheatham, S.Foam Rolling Prescription: A Clinical Commentary. *J Strength Cond Res 34*: 3301–3308, 2020.
3. Behm, D.G., Blazevich, A.J., Kay, A.D., and McHugh, M.Acute effects of muscle stretching on physical performance, range of motion, and injury incidence in healthy active individuals: a systematic review. *Appl Physiol Nutr Metab 41*: 1–11, 2016.
4. Behm, D.G. and Chaouachi, A.A review of the acute effects of static and dynamic stretching on performance. *Eur J Appl Physiol 111*: 2633–2651, 2011.

5. Behm, D.G., Kay, A.D., Trajano G.S., and Blazevich, A.J.Mechanisms underlying performance impairments following prolonged static stretching without a comprehensive warm-up. *Eur J Appl Physiol* 121: 67–94, 2021.
6. Chaabene, H., Behm, D.G., Negra, Y., and Granacher, U.Acute Effects of Static Stretching on Muscle Strength and Power: An Attempt to Clarify Previous Caveats. *Front Physiol* 10: 1468, 2019.
7. Dintiman, G. and Ward, B. *Sport Speed.* Windsor, ON: Human Kinetics, 2003.

2 History of Stretching

Stretching has been and continues to be a controversial training technique. When did it begin? Is it actually beneficial? Often, I have heard individuals state that if stretching was important for performance enhancement, you would see lions, tigers, and cheetahs meeting in groups (wearing spandex and knee braces) for a stretching routine before going on a hunt. While our mammalian cousins are not quite that organized, we do see animals stretching after sleeping or lounging for an extended period of time. This ritualized behaviour is called pandiculation. Pandiculation involves a voluntary contraction of the muscles, followed by a slow stretch/elongation and then relaxation (often with a yawn) (see Figure 2.1). The description is somewhat similar to contract-relax proprioceptive neuromuscular facilitation (PNF) (without the yawn), which was reported to be developed by Herman Kabat, but perhaps he stole the idea from his pet dog or cat (if he had a pet?). One simply has to watch a pet dog or cat and see them stretching their fore and

Figure 2.1 Pandiculation: involuntary stretching of soft tissues, with most mammals associated with transitions between cyclic biological behaviours, especially the sleep-wake rhythm

DOI: 10.4324/9781032709086-3

hind limbs after napping. Animals stretch (pandiculate) all the time! Is it a reflex ritual that actually benefits performance (hunting prey or escaping from predators)? Lions and tigers are not very accessible or easy subjects to recruit and I do not know of any stretch training studies using domesticated pets. Thus, there is no evidence for or against the effectiveness of pandiculation for performance enhancement. Obviously, animals (other than the human animal) have not organized and categorized their stretching and warm-up activities as humans have. Neither do animals stretch for such long durations or intensely as many human athletes or trained individuals. Of course, they also do not play tennis for 3–4 hours like professional tennis players, crash into each other for more than an hour like American football, rugby, and ice hockey players, perform *multiple* double or triple rotations of their bodies with twists for over 1–3 minutes like gymnasts, dancers, figure skaters, trampolinists, and others. Other mammals can be far more athletic than humans, but they tend to perform their athletic pursuits as a prey or a predator over a brief period, sporadically and without hours of training per day over months and years. We as humans tend to be far more obsessive, and compulsive in our activities and likewise our training for those activities. There is a common assumption that animals never get injured. There is no evidence for that assumption. Racehorses for example, commonly sustain muscle strains and have a high incidence of tendinopathy (23). Dogs suffer strains (muscles) and sprains (ligaments) (18). So maybe if they had better warm-up and stretching routines, they would experience less injuries? It is also argued that animals and ourselves often sprint without any warm-up (e.g., we are late for a bus!) and many times either catch the prey, escape from the predator or actually catch the bus. Thus, the contrarians argue there is no need for a warm-up. The problem with this argument is there is no control group! Nobody has taken a pride of lions and systematically stretched them over time (experimental group) and compared their performance to a control group of lions that did not stretch or only pandiculated. Perhaps, the lion would be incrementally faster if it did do a systematic warm-up. While, there is no evidence available for wild predators and prey, there is a large body of human research available regarding warm-up-induced positive changes to athletic performance (14), oxidative energy metabolism, increased lactate clearance, increased blood flow due to muscle capillary vasodilation (7), post-activation potentiation, neural conduction velocity, enzymatic cycling, tissue compliance, and a myriad of other contributing physiological factors (4, 5). We will cover many of these topics in the subsequent chapters.

When did humans start stretching in an organized manner with the hope of enhancing performance or decreasing the chance for injury? Although in historical texts, stretching is not always precisely listed as an activity preceding training or competition, it might be logical to assume that military personnel or athletes would have indulged in some kind of "limbering up" exercises to warm the muscles and body and increase the compliance or decrease the stiffness of the muscle and tendons. The martial arts of many Asian countries are well known for their emphasis on extreme range of motion (ROM) in order to perform high kicks, acrobatic, and escape manoeuvres (think of Bruce Lee's athletic ability). The martial arts of the Chinese, Japanese and Aleut peoples, as well as Mongolian wrestling are suggested to have originated in the prehistoric era (11). Chinese boxing has been traced to the Zhou dynasty (1122–255 BCE). Martial arts like Kung Fu were influenced by Indian martial arts that spread to China in the fifth to sixth century CE (11). With the physical contortions, commonly observed with wrestling and martial arts, it would seem likely that the combatants would have needed to work on their flexibility to improve their chances of succeeding.

We are all familiar with the impressive flexibility of experienced yoga practitioners as they move from one difficult posture to another. Asia and specifically the Indus-Sarasvati civilization in Northern India are credited for the origins of Yoga during what is termed the pre-classical yoga period approximately 5,000 years ago and perhaps even earlier. However, this early yoga practice concentrated on the mind and spirit with little to no emphasis on the physical. Breath control exercises – the precursor of yoga – were implemented in China around 2600 BCE. It is not known if stretching exercises were also included but an exercise chart (168 BCE) of breathing and postures was developed for Tao Yin activities in the early Han dynasty. These exercises or postures were purported to cure specific illnesses (16). Perhaps, this is where the idea that stretching decreases injuries first began. Around the fourth century, during what is classified as the post-classical Yoga (classical Yoga period defined by Patanjali's Yoga-Sûtras "eight limbed path" to Samadhi or enlightenment starting in the second century CE), a system of practices was created to rejuvenate the body, prolong life and embrace the physical body to achieve enlightenment. Tantra Yoga, was developed to cleanse the body and mind. The evolution of these physical-spiritual connections and body-centred practices moved towards the development of Hatha Yoga (Figure 2.2).

Figure 2.2 Yoga positions

Stretching is specifically mentioned as an important component of an exercise regimen to prevent illness by Hua Tuo (104–208 CE). Hua Tuo suggested mimicking animal movements such as walking like a bear, and stretching the neck like a bird (even 2,000 years ago people noticed that animals stretched or at least pandiculated) among other animal-like movements. He emphasized combinations of breathing, bending, stretching and an assortment of postures which he labelled as the "Frolics of the Five Animals" (16). Tai Chi may trace its beginnings to these frolicking exercises.

In the Western civilizations, the Greeks held festivals (Tailteann Games: circa 1,800 BCE) that involved stone throwing, jumping, spear throwing, wrestling, and other activities (16). During pharaonic Egyptian times, athletes also wrestled, boxed, swam, ran, and lifted heavy objects in competition. One can imagine, that these athletes had some kind of pre-competition preparation (warm-up) and that especially for sports like wrestling where limbs can be forced and placed in extreme positions, preparatory stretching would have taken place.

Were these early stretching exercises more static or dynamic in nature? First of all, how do we define static and dynamic stretching? Static stretching involves lengthening a muscle until either a stretch sensation or the point of discomfort is reached and then holding the muscle in a lengthened position for a prescribed period of time (1, 2, 3). Dynamic stretching involves the performance of a controlled movement through the ROM of the active joint(s) (2, 3). Both types of stretching have gone through periods of popularity and disfavour. For example, Hua Tuo's stretching the neck like a bird exercise would likely have involved a slow dynamic component to reach the end of the ROM and then a static component to hold that position. Dynamic stretching was popular in Persia where warriors and wrestlers starting around the first century AD used implements shaped like bowling pins called meels (Figure 2.3). While the heavy meels weighed approximately 50 pounds and would have been used for strength and

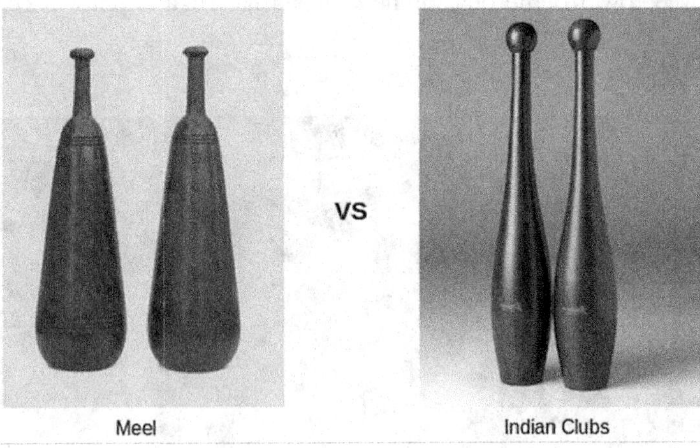

Figure 2.3 Persian meels and Indian clubs

power enhancement, the 2-pound meels were swung in patterns around the shoulder and would have been excellent for a dynamic warm up of the muscles and increasing the ROM. The Persians introduced this form of exercise to the Indian subcontinent in the thirteenth century. The people of the India subcontinent called this activity Persian yoga (16). It is quite likely that similar movement variations with a variety of weapons (i.e., short and long swords) would have been practiced by medieval knights in preparation for combat and competitions. Incorporating the movement of moderate to heavy loads as part of the pre-activity preparation (warm-up) is incorporated today to induce post-activation potentiation enhancement (PAPE) (6). PAPE not only improves subsequent strength and power performance by the aforementioned factors (i.e., neural conduction velocity, muscle vasodilation, increased rate of energy metabolism, etc.) but would also induce myosin light chain phosphorylation, which increases the sensitivity of the myofibrillar proteins (e.g., myosin, actin, tropomyosin, troponin) to calcium thus enhancing force output (4, 5).

Generally, such light dynamic movement for flexibility, quickness and muscular endurance was in the purview of men getting ready for battles or tournaments. However, in the mid-nineteenth century, exercises based on Swedish (Ling's) gymnastics were introduced to women in Europe and North America. They involved the graceful moving of arms, legs, neck, and head. Although their primary purpose (training for family life) would not have been to increase flexibility, such movements would have maintained or enhanced movement around the exercised joints. These *exercise-liberated* women could now be more effective at reaching farther across to make the bed in the mornings, extending further to scrub the floor under the furniture or reach deeper in the cupboard for the pots and pans for making dinner (training for family life!). Quite the liberation! Men, for their part, would still incorporate low intensity or light dynamic movements for sport preparation or war.

The late nineteenth century saw the emergence of a number of new team sports such as ice hockey (March 3, 1875: Montreal, Quebec, Canada), baseball (1672–1700: England), basketball (1891 by James Naismith (Canadian) at Springfield College, Massachusetts, USA), North American football (November 6, 1869, Rutgers vs. Princeton University, USA), volleyball (1895 by William G. Morgan from Springfield College, Massachusetts, USA) and others. The typical competitive zeal of the human athlete would instil a need to find a perceived advantage even in these early days of these new sports. One of those advantages could be a proper warm-up to prepare the body for competition and part of that warm-up would include light dynamic movements for "limbering up".

But more importantly was the defence of your country and during the World Wars, soldiers were systematically trained to ensure they were ready for heavy military action. The systematic training of soldiers was incorporated during the world wars with scientific investigations of optimal resistance training routines sought out by Colonel DeLorme of the United States armed forces (9, 10). Subsequently, in the late 1950s and published in 1961, a Canadian, Dr. William Orban developed the 5BX (5 Basic eXercises) programme (22). Though, it was originally targeted at male military personnel (air force pilots) who might not have access to training equipment and thus could perform calisthenics in most any location to stay fit, it spread to the general population. Orban also developed the XBX which were ten basic exercises modified for women. One of the stated objectives was to "Keep the important muscles and joints of the body supple and flexible." Some of the exercises were quite dynamic and ballistic such as the toe touching exercises which involved bobbing up and down by flexing at the hips to touch the toes and then bounce back to an erect standing position.

However, the appeal of dynamic ballistic stretching activities diminished when it was noted that dynamic stretching of the muscles activated reflexes such as the myotactic (i.e., stretch) reflex, which results in reflexive contractions of the actively stretched muscle (19, 20). If the goal of stretching was to increase ROM, then it was reasoned that dynamic activities that elicited reflex muscle contractions while elongating the muscles could result in injury. Two forces would be working against each other with muscle elongation from stretching opposing stretch reflex-induced contractions. Hence, during the mid-1960s and thereafter, static stretching replaced ballistic or dynamic stretching as the predominant activity within a pre-activity warm-up routine to increase ROM (26, 27). Static stretching was recommended, as the slow movement into the stretch position and maintenance of a static stretch over a prolonged period minimized the reflexive firing of the muscle spindles (primarily the nuclear chain and bag fibres) that were activated by higher rates of stretch (12, 17). Hence, the attenuation of reflex activity with prolonged static stretching would presumably result in a more relaxed muscle and theoretically allow greater muscle lengths to be achieved. Thus, for the next 30 or more years, static stretching was the predominant form of stretching for warm-ups and flexibility.

In the 1970s another stretching technique also became more popular: proprioceptive neuromuscular acilitation (PNF) stretching (8). PNF was developed around 1946 by Herman Kabat, a neurophysiologist. The techniques evolved over time, but one popular variation of PNF was the contract-relax-agonist-contract (CRAC) method. If you wanted to stretch the hamstrings you would CONTRACT the hip flexors (i.e., quadriceps) till you reached your maximum ROM. Then you would RELAX as your partner held that elongated position, which would be followed by CONTRACTION of the hamstrings (AGONISTS). These variations of PNF were purported to induce a number of inhibitory reflex mechanisms (e.g., depressed H-reflex, reciprocal inhibition, autogenic inhibition: 21, 25) that would relax the muscle allowing the individual to reach even greater increases in ROM than with static stretching (8). You could not go to any team sport in the 1970s without seeing athletes pairing up to passively stretch and provide resistance to their partner's contractions of elongated muscles. However, not many people questioned at the time, why a technique like PNF that supposedly inhibited excitatory reflexes would be used in a warm-up that should excite the system in preparation for high intensity activity.

In the late 1990s and early 2000s scientific reports began to appear indicating that static stretching rather than enhancing subsequent performance might actually impair performance (13, 24). As the evidence began to mount throughout the early 21st century, static stretching was replaced with dynamic stretching as the major flexibility component of the warm-up. Only recently, has the evidence for static stretch-induced performance impairments been suggested to lack some ecological validity (practical reality or real-life application). A position stand/review by the Canadian Society for Exercise Physiology (2) published in 2016 documented that many of the static stretching studies did not employ a prior aerobic-style warm-up, stretched the muscle(s) for durations much longer than are typically used, did not include any dynamic sport specific activities after stretching and conducted the testing within 3–4 minutes of the experimental protocol. Another study found that just the knowledge of the previously published stretching impairment studies (expectancy bias) could negatively affect the results (15). Thus, the state of stretching within a warm-up and for improving ROM is in a state of flux and confusion. The objective of this book is to alleviate some of that confusion by critically analyzing the 5Ws of stretching with an "H" thrown in for good measure. That is, 1) **What** are the effects and physiological mechanisms underlying different types of stretching? 2) **Why** should we stretch? 3) **When** should we stretch? 4) **Who** are the major pioneers, innovators and researchers in this area? 5) **Where** does the science of flexibility and stretching go next and 6) **How** should we stretch or use other techniques to increase ROM?

Summary

Many animals elongate (stretch) their muscles after a period of rest. Although, these animals have pandiculated for eons, it is not known whether there are any performance benefits. There is inferential evidence that humans have stretched for thousands of years with the advent of Yoga (~3000 BCE) and martial arts (~1000 BCE) in Asia. The Greeks and Egyptians probably emphasized dynamic actions and stretches prior to their athletic competitions (~2000 BCE). Static stretching became more popular after the World Wars, while PNF stretching was popularized to a greater degree in the 1970s. Both stretching styles remained predominant until the 1990s, when research began to appear indicating that static stretching could lead to performance impairments. Since that time dynamic stretching has made a resurgence. The most recent studies suggest that the move away from static stretching may have been premature and based on impractical study designs.

References

1. Behm, D.G., Bambury, A., Cahill, F., and Power, K.Effect of acute static stretching on force, balance, reaction time, and movement time. *Med Sci Sports Exerc* 36: 1397–1402, 2004.

2. Behm, D.G., Blazevich, A.J., Kay, A.D., and McHugh, M.Acute effects of muscle stretching on physical performance, range of motion, and injury incidence in healthy active individuals: a systematic review. *Appl Physiol Nutr Metab* 41: 1–11, 2016.
3. Behm, D.G. and Chaouachi, A.A review of the acute effects of static and dynamic stretching on performance. *Eur J Appl Physiol* 111: 2633–2651, 2011.
4. Bishop, D.Warm up I: potential mechanisms and the effects of passive warm up on exercise performance. *Sports Med* 33: 439–454, 2003.
5. Bishop, D.Warm up II: performance changes following active warm up and how to structure the warm up. *Sports Med – ADIS Int* 33: 483–498, 2003.
6. Blazevich, A.J. and Babault, N.Post-activation Potentiation Versus Post-activation Performance Enhancement in Humans: Historical Perspective, Underlying Mechanisms, and Current Issues. *Front Physiol* 10: 1359, 2019.
7. Brunner-Ziegler, S., Strasser, B., and Haber, P.Comparison of metabolic and biomechanic responses to active vs. passive warm-up procedures before physical exercise. *Journal of Strength and Conditioning Research / National Strength & Conditioning Association* 25: 909–914, 2011.
8. Burke, D.G., Culligan, C.J., and Holt, L.E.The theoretical basis of proprioceptive neuromuscular facilitation. *Journal of Strength and Conditioning Research* 14: 496–500, 2000.
9. Delorme, T.Restoration of muscle power by heavy-resistance exercises. *The Journal of Bone and Joint Surgery* 27: 645–667, 1945.
10. Delorme T., Ferris, B., and Gallagher, J.Effect of progressive resistance exercise on muscle contraction time. *Arch Phys Med* 33: 86–92, 1952.
11. Draeger, D.S.J. *Comprehensive Asian Fighting Arts*, 1969.
12. Durbaba, R., Taylor, A., Ellaway, P.H., and Rawlinson, S.The influence of bag2 and chain intrafusal muscle fibers on secondary spindle afferents in the cat. *J Physiol* 550: 263–278, 2003.
13. Fowles, J.R., Sale, D.G., and MacDougall, J.D.Reduced strength after passive stretch of the human plantar flexors. *Journal of applied physiology* 89: 1179–1188, 2000.
14. Fradkin, A.J., Zazryn, T.R., and Smoliga, J.M.Effects of warming-up on physical performance: a systematic review with meta-analysis. *J Strength Cond Res* 24: 140–148, 2010.
15. Janes, W.C., Snow, B.B., Watkins, C.E., Noseworthy, E.A., Reid, J.C., and Behm, D.G.Effect of participants' static stretching knowledge or deception on the responses to prolonged stretching. *Appl Physiol Nutr Metab* 41: 1052–1056, 2016.
16. Kunitz, D. *Lift: Fitness Culture from Naked Greeks and Acrobats to Jazzercize and Ninja Warriors.* New York: Harper Wave, 2016.
17. Laporte, Y., Emonet-Dénand F., and Jami, L.The skeletofusimotor or ·-innervation of mammalian muscle spindles. In: *The Motor System in Neurobiology*, E.V. Evarts, S.P. Wise, and D. Bousfield (Eds). New York: Elsevier Biomedical Press, 1985, pp. 173–177.
18. Levy, M., Hall, C., Trentacosta, N., and Percival, M.A preliminary retrospective survey of injuries occurring in dogs participating in canine agility. *Vet Comp Orthop Traumatol* 22: 321–324, 2009.
19. Matthews, P.B.C.Developing views on the muscle spindle. In: *Spinal and Supraspinal Mechanisms of Voluntary Motor Control and Locomotion*. 1980, pp. 12–27.
20. Matthews, P.B.C.Muscle spindles: their messages and their fusimotor supply. In: *The Nervous System: Handbook Of Physiology.* V.B. Brooks (ed.). American Physiological Society, 1981, pp. 189–288.

21. Moore, M.A. and Kukulka, C.G.Depression of Hoffmann reflexes following voluntary contraction and implications for proprioceptive neuromuscular facilitation therapy. *Physical Therapy* 71: 321–329, 1991.
22. Orban, W.A.R. *Royal Canadian Air Force Exercise Plans for Physical Fitness.* Ottawa, ON: Queen's Printer, 1962.
23. Parkin, T.D.Epidemiology of racetrack injuries in racehorses. *Vet Clin North Am Equine Pract* 24: 1–19, 2008.
24. Power, K., Behm, D., Cahill, F., Carroll, M., and Young, W.An acute bout of static stretching: effects on force and jumping performance. *Med Sci Sports Exerc* 36: 1389–1396, 2004.
25. Sharman, M.J., Cresswell, A.G., and Riek, S.Proprioceptive neuromuscular facilitation stretching: mechanisms and clinical implications. *Sports Med* 36: 929–939, 2006.
26. Young, W.B.The use of static stretching in warm-up for training and competition. *Int J Sports Physiol Perform* 2: 212–216, 2007.
27. Young, W.B. and Behm, D.G.Effects of running, static stretching and practice jumps on explosive force production and jumping performance. *J Sports Med Phys Fitness* 43: 21–27, 2003.

3 Types of Stretching and the Effects on Flexibility

There are a number of different types of stretching and the public can be confused regarding their differences. Passive and active static stretching, dynamic, ballistic, proprioceptive neuromuscular facilitation (PNF), and others are used to enhance flexibility or range of motion (ROM) as well as be incorporated as part of a pre-competition or training warm-up. First of all, what is the definition of flexibility? Michael J. Alter in his textbook (6) lists a variety of definitions. Some of the definitions of flexibility include:

1 ROM available to a joint or group of joints (50, 85, 93, 116, 185)
2 Total achievable excursion (within limits of pain) of a body part through its potential ROM (167)
3 Ability to move a joint smoothly and easily through its complete pain free ROM (100, 101)
4 Ability to move a joint through a normal ROM without undue stress to the musculotendinous unit (41)
5 Normal joint and soft tissue ROM in response to active or passive stretch (81)

The ability to increase a joint's ROM, would normally necessitate an improved extensibility (78) or decreased stiffness of musculotendinous and other connective tissues. Gajdosik and colleagues (65, 66) suggest that flexibility should be described as a ratio of change in muscle length or joint angle to a change in force or torque. Recent research use this description or technique by subjecting a limb joint to an extended ROM on an isokinetic dynamometer as a test for muscle stiffness (122, 124, 136, 139, 146).

To achieve these increases in flexibility, a variety of stretching techniques are used. Static stretching for example, involves lengthening a muscle until either a stretch sensation or the point of discomfort is reached and then holding the muscle in a lengthened position for a prescribed period of time (23, 24, 26, 48). Whether it is passive or active static stretching depends upon whether the muscle is lengthened by an external force (i.e., another person, or a tool like a stretching band or machine) with the muscle relaxed (passive static stretch) or lengthened by an active contraction of the affected muscle or other muscles (i.e., antagonist) (active static stretch) (see Figure 3.1). Static stretching is used in athletic, fitness,

DOI: 10.4324/9781032709086-4

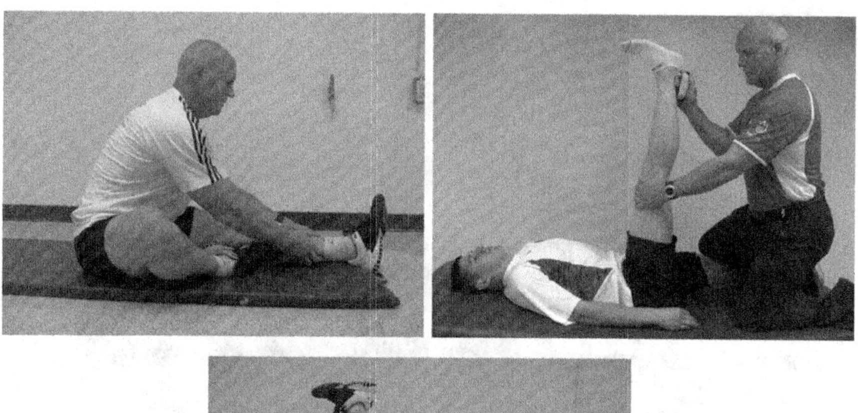

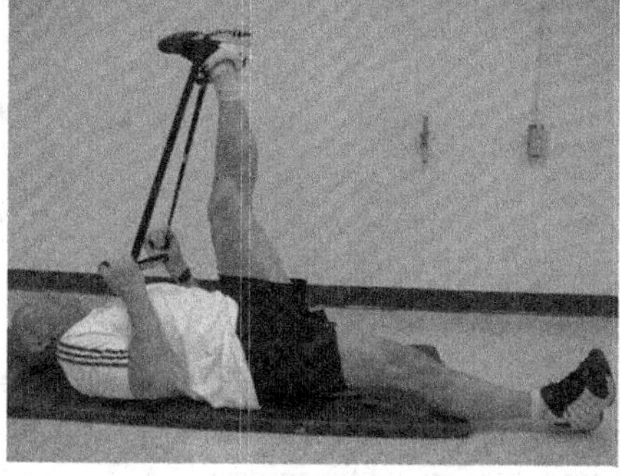

Figure 3.1 Unassisted, and assisted passive static stretching with a partner or a band

health and rehabilitation environments. It is an effective method for increasing joint ROM (14, 15, 162) and was purported to improve performance (198, 199) and reduce the incidence of activity-related injuries (23, 24, 26, 126, 169, 170, 182) (primarily musculotendinous injuries associated with explosive and change of direction actions, 24, 29). However, the possibility of static stretch-induced performance impairments has limited its use in the new millennium. Evidence for and against this bias will be presented in a subsequent chapter.

Dynamic stretching uses a controlled movement through the ROM of the active joint(s) (61). It can be exemplified by swinging the legs back and forth (hip flexion and extension) or side to side (hip abduction and adduction) or swinging the arms in circles (shoulder circumduction) (see Figure 3.2). Dynamic stretching differs from ballistic stretching in that ballistic stretching would typically involve higher velocity movements with bouncing actions at the end of the ROM (10, 148). Ballistic movements were used in the aforementioned 5BX programme (popular from 1950s until the 1960s) and were still prevalent till quite recently in many military and police style training. In the late 1980s, I consulted with fitness instructors of the Royal Canadian Mounted Police at their national training depot in Regina, Saskatchewan, Canada. Recruits were trained and tested by performing resistance exercises such as shoulder presses, push-ups, sit-ups and others as quickly as possible in a prescribed time. The number of repetitions completed in 30 or 60 seconds

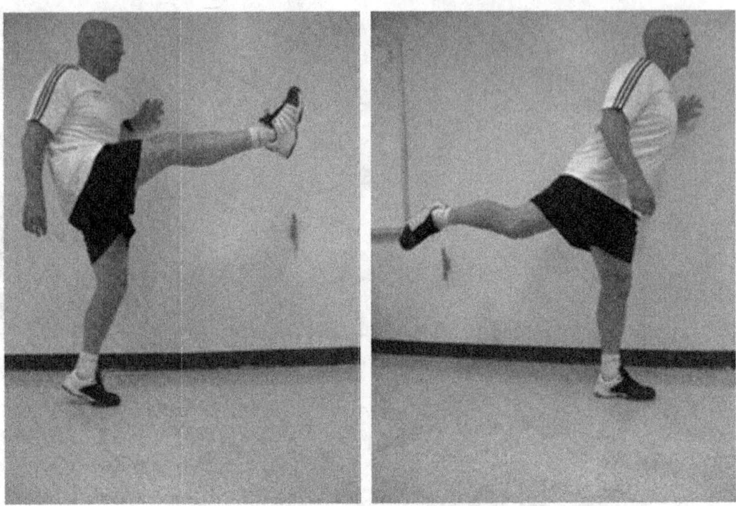

Figure 3.2 Dynamic stretching (hip extension and flexion)

was measured and thus the recruits would ballistically throw the barbells up and down (i.e., shoulder press) or slam their trunk back and forth (i.e., sit-ups) as quickly as possible. Definitely a recipe for injuries!

What is the difference between dynamic stretching and dynamic activity? It could be argued that dynamic stretching is a dynamic activity but not all dynamic activities are considered dynamic stretching. The decisive factor is whether the dynamic activity moves the body through a full or nearly full ROM. Jogging, skipping, hopping and other similar activities are all dynamic, but as they only emphasize a restricted or small to moderate ROM they would not be considered as dynamic stretching. However, if the person did butt (gluteal) kicks (knee flexion and touches the buttocks with heel of the foot) while jogging, then this dynamic activity would be under the purview of dynamic stretching as it goes through a fuller ROM. As mentioned in the historical section, dynamic stretching in this millennium was considered preferable to static stretching in a warm-up, owing to training specificity (training movement matches the sport or exercise movements) (28, 172) as well as activity-induced increases in metabolism, muscle temperature (30, 62, 197, 198), and neural activation (73, 74, 75). Many systematic warm-up routines such as the FIFA 11 (football/soccer specific warm-up) utilize a combination of dynamic stretching and activities in preparation for subsequent activity and with evidence for decreased injury incidence (1, 2, 82, 149, 155, 180, 181).

PNF stretching combines static stretching and isometric contractions in a cyclical pattern. PNF was developed in the late 1940s and early 1950s by Herman Kabat, and two physical therapists, Margaret Knott and Dorothy Voss. Kabat, was a neurophysiologist, who developed PNF based on the neuromuscular research of Sir Charles Sherrington (178, 179). Two of the more ubiquitous techniques are the contract relax (CR) and contract-relax-agonist-contract (CRAC) techniques (168,

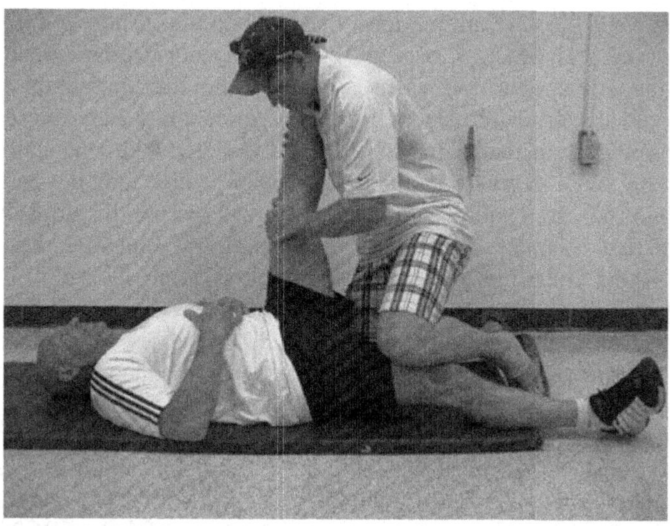

Figure 3.3 Proprioceptive Neuromuscular Facilitation (PNF) stretching for the hamstrings

177). The CR method includes a static stretching component followed immediately by an isometric contraction of the stretched muscle, followed with another stretch of the target muscle. CRAC involves an additional contraction of the agonist muscle (i.e., opposing the muscle group being stretched) during the stretch, prior to the additional stretching of the target muscle (see Figure 3.3). PNF was used extensively by team athletes (PNF stretches typically need a partner) in the late 20th century, but similar to static stretching, its use has diminished in this millennium. A number of individual studies suggest that PNF is more effective than static or dynamic stretching for improving ROM (70, 87, 154). However, a recent meta-analysis systematic review reported an advantage for static stretch training over PNF training for increasing ROM (187).

Why should we stretch? Stretching is primarily used to increase joint ROM. When used as part of a warm-up, the increased ROM was thought to improve performance and decrease the incidence of injuries (24, 26, 197). Although, the ability to increase ROM through stretching is generally universally agreed, its impact on performance and injuries is more controversial. There are many factors to consider including the type, duration (volume), intensity of stretching, the population that is stretching (i.e., male, female, young, old, athletic, sedentary and others) among other factors. Others may use stretching to achieve a greater sense of relaxation (i.e., yoga) (psychological or neurological effects).

Stretching-Induced Changes in Range of Motion

First of all, flexibility is not a global phenomenon of the body (7) but stretching effects can be global (20, 22). That is, an individual can be flexible in one joint but not others (7). Even within a joint they may have a greater ROM with one movement versus another. A baseball pitcher may not exhibit similar relative flexibility as a discus thrower when comparing shoulder horizontal extension versus shoulder internal and external rotation. Some joints are more susceptible to stretch-induced increases in ROM. For example, stretching calf muscles provides only small increases in ankle dorsiflexion, which may not be clinically meaningful (164). In a subsequent chapter, we will discuss the global effects of stretching; that is: unilaterally stretch one muscle or joint and there can be increased ROM in another non-stretched muscle or joint.

What factors restrict our joint flexibility? Muscle fascicles (structural proteins like myosin, actin, titin and others), tendons, aponeuroses, joint capsules, and ligaments contribute to ROM restrictions with passive muscle elongation. As the composition of a human can range from 50–78 percent fluid (water) dependent on your age, sex, hydration, and other factors, the viscosity of the tissues can substantially affect flexibility. Furthermore, a highly activated central nervous system could increase muscle tonus and with a less relaxed muscle inhibit flexibility. In addition, pushing a joint to its maximum ROM can be uncomfortable or painful and thus the ability to tolerate this discomfort or pain may allow some individuals to stretch farther than other more pain sensitive individuals (121).

Stretching can induce elastic or plastic changes of the musculotendinous system. Elastic changes are defined as the elongation of tissues that recovers when the tension is removed; so, it is a temporary increase in ROM. Plastic changes involve a musculotendinous elongation where the tissue deformation remains even after the tension is removed; so, it is considered a semi-permanent change in flexibility. Elastic changes are the increases in ROM that are experienced immediately after a single session of stretching. A single session of static stretching can induce small-to-large relative ROM increases that persist for five (193), ten (23, 25, 27), 30 (63), 90 (102), and 120 minutes (162). Evidence of persistent improved ROM even one day later has been reported (137). However, not all studies show prolonged ROM increases. Two 15-second passive stretches did not provide a significant improvement in ROM at any time post-stretch (200), whereas augmented ROM was only evident for 3 minutes following four 30-second static stretches (52) and only for six minutes following five-modified hold and relax stretches (184). Specific training-induced plastic changes and mechanisms will be discussed in a later section.

As mentioned previously, there is much conflict as to what type of stretching provides the greatest ROM. PNF has been reported to provide greater acute or elastic ROM improvements than static stretching in some studies (57, 60, 151), which contrasts with other studies finding similar ROM changes between PNF and static stretching (46, 120). Dynamic stretching is also controversial with some studies reporting that a single session of dynamic stretching provides either similar (19, 158) or greater (8, 56) ROM increases as static stretching. But once again, conflicting studies report that dynamic stretching is not as effective as static stretching when used as part of warm-up

(10, 16, 156, 174, 176). When comparing static to ballistic stretching in a single (acute) stretching session, ballistic stretching has been reported to provide less ROM than static (10), or PNF (16), but similar ROM as static stretching and PNF in another study (104). Typically, we rely on systematic meta-analytical reviews to clarify the confusing messages from the wide variety of findings from small sample original research studies. A recent meta-analysis (21) of 47 studies revealed that a single session of stretching on average provided a small magnitude increase in ROM compared to non-active (non-stretched) control conditions. Whereas there were ROM increases with sit and reach, hamstrings, and triceps surae tests, there was no change with the hip adductor flexibility. It was also noted that stretch intensity, trained state, stretching techniques, or sex did not significantly modulate the enhanced ROM. A meta-regression demonstrated no relationship between the ROM increases with age or stretch duration.

In contrast to acute responses, is there a different response when examining chronic training study results? Another recent meta-analysis by our group (103) of 77 stretch training studies reported that chronic stretch training effects induced moderate magnitude ROM increases. With stretch training (in contrast to acute stretching effects) there were significant differences between the stretching techniques, with PNF and static stretching exhibiting greater increases in ROM than ballistic/dynamic stretching. Furthermore, females showed greater ROM training gains compared to males, whereas stretch volume, intensity, or frequency did not play a significant role in ROM gains. A 2018 meta-analysis systematic review determined only trivial differences in ROM improvements between acute and chronic PNF and static stretching (127), whereas another 2018 review found that chronic static stretch training provided greater flexibility than PNF training (187). Ballistic stretching (higher velocity movements with bouncing actions at the end of the ROM), which can be considered a variation of dynamic stretching was not as effective as static stretching for improving ROM after a 4-week training programme (47). Only one study compared PNF and dynamic stretching and indicated that after 14 training sessions that PNF provided 3–7 percent greater ROM increases (192).

Another form of stretching is termed neuromobilization. With neuromobilization, the nerve is placed under a lengthening stress. For example, when performing a seated straight leg raise, the individual would actively dorsiflex the foot and flex the cervical spine to exert a type of neural traction (Figure 3.4). In an 8 week stretch training study, neuromobilization provided a greater increase in hamstring length versus PNF in the first 4 weeks but in the last 4 weeks, hamstrings length actually decreased. Thus, overall, passive static stretching provided the greatest benefits (58).

Our psychological state or emotions can affect flexibility. Two sessions of 20 minutes each of anti-anxiety techniques such as the neuro-emotional technique, (also known as a mind body technique) was shown in one study to enhance ROM to a greater extent than a similar duration of passive static stretching or no stretching (96). It is purported that latent anxiety, which even

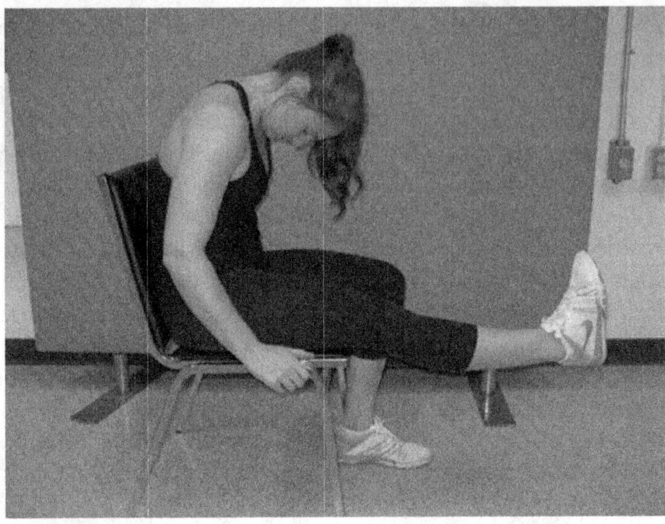

Figure 3.4 Neuromobilization

if not presently apparent can lead to learned emotional responses that affect the motor responses. Anti-anxiety techniques would dissipate these responses and lead to greater muscle relaxation and lower muscle tonus. Various forms of yoga combine the physical aspects of stretching with meditation or relaxation techniques to enhance the effects on flexibility.

Yoga

Serious yoga practitioners are widely known for their enhanced flexibility (161). However, yoga is also reported to improve a wide array of physiological and psychological parameters. For instance, it has been reported to increase muscular strength (119, 189) and endurance (189), balance (161), maximal oxygen uptake (189), improved reaction times (119), breath holding times (119), reduced cardiovascular risk, blood pressure, body mass index (13), unify the body, mind, and spirit, among others. How can yoga stretching accomplish all these beneficial measures? Yoga misperception is the problem. Yoga is not just stretching but involves many other components. In addition to the stretching aspect, full yoga practitioners should experience changes in their mental attitude (i.e., meditation), diet, practice of specific techniques for postures (*asanas*) and breathing practices (*pranayamas*) in order to attain a higher level of consciousness. It is unlikely that the average North American or European who goes to a trendy hot yoga session or most yoga sessions is serious or committed to a degree that they will attain a new level of consciousness. However, 30–60 minutes of stretching, changing positions and holding various postures (dynamic and static muscle endurance) and inspiratory and expiratory breathing techniques will certainly enhance a variety of physiological and health parameters. From personal experience, having only participating in a single 60-minute yoga session, I experienced exercise-induced muscle soreness and delayed onset muscle soreness for days thereafter. Although I am considered a relatively very fit individual for my age, I was not accustomed to the prolonged and slow eccentric contractions when moving from one posture to the next or the ability to hold certain postures under extended muscle positions for prolonged periods. A lack of task specificity was certainly evident in my muscle pain over the next few days. Hence, you should not directly compare yoga to stretching practices like static dynamic, ballistic and PNF, as stretching is only one component of yoga. Furthermore, there are many types of yoga that may place greater emphasis on holding positions for longer durations to achieve changes in connective tissue (e.g., yin yoga) or place greater emphasis on breathing patterns or meditation.

Unfortunately, science is messy and results are not always consistent. The same interventions and measures can be used on sedentary women and men, youth, young adult and seniors, trained and untrained and you can get different results in every study. How do you figure out what is right? How do you know what to do? As mentioned, meta-analyses are usually considered the gold standard for integrating all the disparate information in the literature into a cogent understandable main message. But the problem with meta-analyses is that sometimes, they might hide some intricacies. Maybe, for the general population of sedentary and recreationally trained people, static and PNF stretching provide the greatest ROM increases. But as gymnasts and figure skaters and circus acrobats are the extremes of the population maybe just for them ballistic stretching is more important, as it emulates the actions in their sport. I just used those extreme athletes as a fictional example. There is very little research on such a small population of extreme flexibility athletes. If a meta-analysis looked at 30, 50, or 100 papers and only 2–3 papers dealt with such highly trained athletes then their responses to specific stretching could be hidden within the greater numbers of the other studies. A good meta-analysis should highlight these outliers but not all reviews accomplish it.

Thus, based on the latest information, it would seem that there is very little difference between stretching techniques when examining a single stretching session but with repetitive stretch training, PNF and static stretching might provide greater increases in ROM than dynamic stretching. However, we will examine later how prolonged static and PNF stretching without a full warm-up protocol might impair subsequent performance. Thus, you need to ask yourself, whether you need the utmost ROM. If you are jogging, the amplitude of your stride length is limited and there is no need to have an extreme ROM. Hence, some dynamic stretching before the running might be sufficient. If you are resistance training, and for your squat warm-up you take a light weight or just your body mass and go through a full ROM for 10 repetitions then that might be sufficient as you will not exceed that range while lifting. A study by Morton et al. (141) compared five weeks of resistance training to static stretch training and found improved ROM in both groups with no significant differences between the groups in the flexibility of the hamstrings, hip flexors, extensors or should extensors. Thus, the ROM with resistance training was sufficient to provide a similar flexibility training adaptation as a stretching programme. We will delve into this topic in more detail later as one of our recent meta-analyses has shown that resistance training can provide similar increases in ROM as static stretching (3). Another study incorporated dancers who either resistance trained, stretched at a low intensity (3/10) or a moderate to high intensity (8/10) for six weeks (195). All groups improved their passive ROM with no difference between groups, whereas the resistance trained and low intensity stretch training groups improved their active ROM. The authors suggested that dance instructors and coaches should incorporate stretching and end-of-range resistance training within their schedules. Stretching would be incorporated at the end of the recovery session. They also recommended that the position of the stretches is very important in order to eliminate muscle contractions and hence the body should be in a stable position without extraneous tension. Each stretch according to this study should be held for 60 seconds at an intensity of 3/10.

The warm-up is not a time to try to make plastic (semi-permanent) changes in your flexibility. The warm-up prepares you for the upcoming activity at hand. There is no need to be able to do the Russian splits (legs completely abducted until the legs are horizontal on the floor) before a 5-km jog or step-ups in the weight room. However, if you have back problems because you spend eight hours per day sitting at a computer and now your pelvis has an anterior tilt, owing to shortened hip flexors affecting your lumbar spinal curvature, then the use of static or PNF stretching as a separate flexibility workout might be in order.

Range of Motion Norms

The average passive joint ROM have been provided in a few studies. The following tables provides a comparison of a sample of these studies in healthy or normal individuals.

Measuring Range of Motion

There are a myriad of instruments that can be used to measure ROM. Most individuals do not have access to advanced scientific laboratories. Hence, the universal, full circle goniometer is one of the most preferred pieces of equipment for measuring ROM (65) (see Figure 3.5). The use of goniometers has been evaluated as having high reliability (intraclass correlation coefficients of >0.91(65). While instruments can have strong time to time (intra-rater) reliability they may not have good between instrument reliability. For instance, a universal, fluid and electronic goniometer were tested for reliability (71).

Table 3.1 Upper Limb and Back Passive ROM in degrees

Joint	*Motion*	*Family Medical Practice*	*Heyward 2005 (86)*
Shoulder	Flexion	180	150–180
	Extension	45–60	50–60
	Abduction	150	180
	External Rotation	90	90
	Internal Rotation	70–90	70–90
Elbow	Flexion		140–150
	Extension		0
Radioulnar	Pronation		80
	Supination		80
Wrist	Flexion		60–80
	Extension		60–70
	Radial Deviation		20
	Ulnar Deviation		30
Cervical spine	Flexion		45–60
	Extension		45–75
	Lateral Flexion		45
	Rotation		60–80
Thoraco-Lumbar Spine	Flexion		60–80
	Extension		20–30
	Lateral Flexion		25–35
	Rotation		30–45

Source: Family Medical Practice: (www.fpnotebook.com/Ortho/Exam/ShldrRngOfMtn.htm).

While the inter-tester reliability in Goodwin's study was excellent (r = 0.90–0.93), the reliability scores between instruments were not as consistent.

Fluid Goniometer vs. Universal Goniometer 0.90
Fluid Goniometer vs. Electrogoniometer 0.33
Universal Goniometer vs. Electrogoniometer 0.51

Another study compared a standard plastic goniometer to a fleximeter (gravitation-based ROM measuring device) or inclinometer (see Figure 3.6). The fleximeter/inclinometer demonstrated moderate to excellent intra- and inter-rater reliability but the goniometer showed poor to moderate intra- and inter-rater reliability (43). These findings indicate that if just one device is used to

Table 3.2 Lower Body Passive ROM in degrees

Joint	*Motion*	*Roass and Andersson 1982 (165)*	*AAOS 1969 (186)*	*Boone and Azen 1979 (33)*	*Heyward 2005 (86)*	*Hallaceli et al. 2014 (80)*
Hip	Extension	9.5	28	12.1	30	19.8
	Flexion	120.4	113	121.3	100–120	128.8
	Abduction	38.8	48	40.5	40–45	45.7
	Adduction	30.5	31	25.6	20–30	24.2
	Internal Rotation	32.6	35	44.4	40–45	43.4
	External Rotation	33.7	48	44.2	45–50	41.9
Knee	Extension		10		0–10	7.53
	Flexion	143.8	134	141	135–150	142.4
Ankle	Extension (Dorsi-flexion)	15.3	18	12.2	20	22.5
	Flexion (Plantar flexion)	39.7	48	54.3	40–50	49.99
	Valgus (Eversion)	27.9	18	19.2	15–20	19.9
	Varus (Inversion)	27.8	33	36.2	30–35	34.1

Source: American Association of Orthopaedic Surgeons.

measure differences before and after stretching or training, the extent of change should be reliable but you cannot always interchangeably use these devices to monitor flexibility differences. Similar conclusions were made when comparing goniometers and a digital level (143). Whereas, intra-tester reliability ranged from 0.91–0.99, inter-tester reliability ranged from 0.31 to 0.95 with limits 2.3 times higher for inter-tester reliability when testing for various shoulder movements (external and internal rotation and flexion). The authors indicated that experienced individuals using the same instrument for repeated measures (goniometer or digital level) should be able to detect a shoulder ROM change of at least 6 degrees, but when comparing measures from two people, the detectable change is 15 degrees. Another study (84) examined the reliability of measuring ROM with visual estimation, goniometry, still photography, "stand and reach" and hand behind back reach for six different shoulder movements. In general, they reported fair to good reliability (r = 0.53–0.73) for visual estimation, goniometry, still photography and stand and reach. However, the tests had standard errors of measurement between 14–25 degrees (inter-rater trial) and 11–23 degrees (intra-rater trial). The hand

Table 3.3 Norms for Joint Range of Motion (°)

	MALES					*FEMALES*				
Joint Movement	*Low*	*Mod. Low*	*Avg.*	*Mod. High*	*High*	*Low*	*Mod. Low*	*Avg.*	*Mod. High*	*High*
	Neck									
Flexion/ extension	<107	107–128	129–142	143–160	>160	<125	125–141	142–160	161–177	>177
Lateral Flexion	<74	74–89	90–106	107–122	>122	<84	84–99	100–116	117–132	>132
Rotation	<141	141–160	161–181	182–210	>201	<158	158–177	178–198	199–218	>218
	Shoulder									
Flexion/ Extension	<207	207–223	224–242	243–259	>259	<226	226–242	243–261	262–278	>278
Adduction/ Abduction	<158	158–171	172–186	187–200	>200	<167	167–180	181–195	196–209	>209
Rotation	<154	154–171	172–192	193–210	>210	<289	189–206	207–227	228–245	>245
	Elbow									
Flexion/ Extension	<133	133–143	144–156	157–167	>167	<133	133–143	144–156	157–167	>167
	Forearm									
Supination/ Pronation	<151	151–170	171–191	192–211	>211	<160	160–179	180–200	201–220	>220
	Wrist									
Flexion/ Extension	<112	112–131	132–152	153–172	>172	<136	136–155	156–176	177–196	>196
Ulnar/radial deviation	<64	64–77	78–92	93–105	>105	<75	75–88	89–101	102–117	>117

Table 3.3 (Cont.)

	MALES					FEMALES				
Joint Movement	*Low*	*Mod. Low*	*Avg.*	*Mod. High*	*High*	*Low*	*Mod. Low*	*Avg.*	*Mod. High*	*High*
	Hip									
Flexion/ Extension	<50	50–67	68–88	89–106	>106	<82	82–99	100–120	121–138	>138
Adduction/ Abduction	<41	41–50	51–61	62–71	>71	<45	45–54	55–65	66–75	>75
Rotation	<59	59–78	79–99	100–119	>119	<90	90–109	110–130	131–150	>150
	Knee									
Flexion/ Extension	<122	122–133	134–146	147–157	>157	<134	134–144	145–157	158–168	>168
	Ankle									
Plantar flexion/ dorsiflexion	<48	48–58	59–71	72–82	>82	<56	56–66	67–79	80–90	>90
Inversion/ Eversion	<30	30–41	42–56	57–68	>68	<39	39–50	51–65	66–77	>77
	Trunk									
Flexion/ Extension	<45	45–62	63–83	84–101	>101	<30	30–47	48–68	69–86	>86
Lateral Flexion	<75	74–89	90–106	107–122	>122	<104	104–199	120–136	137–152	>152
Rotation	<108	108–126	127–147	148–166	>166	<134	134–152	153–173	174–192	>192

Source: Using Leighton Flexometer (88, 114).

Table 3.4 Percentile ranks for the Sit-and-Reach Test (cm) (88)

	AGE (Years)									
	20–29		*30–39*		*40–49*		*50–59*		*60–69*	
% Rank	*M*	*F*	*M*	*F*	*M*	*F*	*M*	*F*	*M*	*F*
90	39	40	37	39	34	37	35	37	32	34
80	35	37	34	36	31	33	29	34	27	31
70	33	35	31	34	27	32	26	32	23	28
60	30	33	29	32	25	30	24	29	21	27
50	28	31	26	30	22	28	22	27	19	25
40	26	29	24	28	20	26	19	26	15	23
30	23	26	21	25	17	23	15	23	13	21
20	20	23	18	22	13	21	12	20	11	20
10	15	19	14	18	9	16	9	16	8	15

Table 3.5 Percentile Ranks for the Modified Sit-and-Reach Test (88)

	FEMALES							
	<18 Years		*19–35 Years*		*36–49 Years*		*>50 Years*	
% Rank	*Inches*	*cm*	*Inches*	*cm*	*Inches*	*cm*	*Inches*	*cm*
99	22.6	57.4	21.0	53.3	19.8	50.3	17.2	43.7
95	19.5	49.5	19.3	49.0	19.2	48.8	15.7	39.9
90	18.7	47.5	17.9	45.5	17.4	44.2	15.0	38.1
80	17.8	45.2	16.7	42.4	16.2	41.1	14.2	36.1
70	16.5	41.9	16.2	41.1	15.2	38.6	13.6	34.5
60	16.0	40.6	15.8	40.1	14.5	36.8	12.3	31.2
50	15.2	38.6	14.8	37.6	13.5	34.3	11.1	28.2
40	14.5	36.8	14.5	36.8	12.8	32.5	10.1	25.7
30	13.7	34.8	13.7	34.8	12.2	31.0	9.2	23.4
20	12.6	32.0	12.6	32.0	11.0	27.9	8.3	21.2
10	11.4	29.0	10.1	25.7	9.7	24.6	7.5	19.1
	MALES							
	<18 Years		*19–35 Years*		*36–49 Years*		*>50 Years*	
% Rank	*Inches*	*cm*	*Inches*	*cm*	*Inches*	*cm*	*Inches*	*cm*
99	20.1	51.1	24.7	62.7	18.9	48.0	16.2	41.1
95	19.6	49.8	18.9	48.0	18.2	46.2	15.8	40.1
90	18.2	46.2	17.2	43.7	16.1	40.9	15.0	38.1
80	17.8	45.2	17.0	43.2	14.6	37.1	13.3	33.8
70	16.0	40.6	15.8	40.1	13.9	35.3	12.3	31.2

Table 3.5 (Cont.)

60	15.2	38.6	15.0	38.1	13.4	34.0	11.5	29.2
50	14.5	36.8	14.4	36.6	12.6	32.0	10.2	25.9
40	14.0	35.6	13.5	34.3	11.6	29.5	9.7	24.6
30	13.4	34.0	13.0	33.0	10.8	27.4	9.3	23.6
20	11.8	30.0	11.6	29.5	9.9	25.1	8.8	22.4
10	9.5	24.1	9.2	23.4	8.3	21.1	7.8	19.8

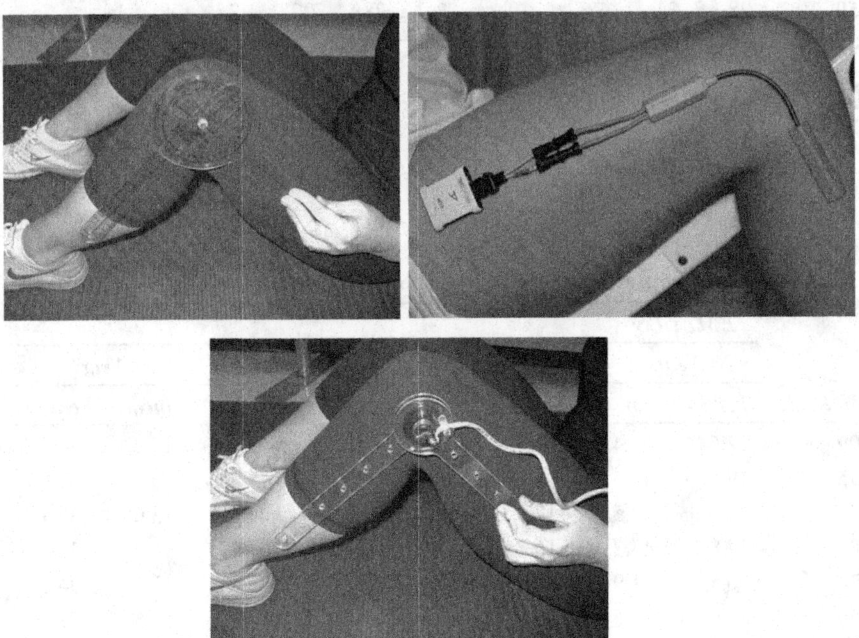

Figure 3.5 Standard and Electronic Goniometers

behind the back test showed poor inter-rater and intra-rater reliability (r=0.14–0.39). These poor findings are probably related to the number of joints involved and the movement complexity. Not all studies recommend visual inspection as van de Pol (190) stated that measurements of physiological ROM using goniometers or inclinometers were more reliable than using vision. However, another review reported that inter-rater reliability of lower extremity passive ROM measurement is generally low (191). One of the major problems is the sensation or measurements of end-feel. End feel is a term used to describe the extent of sensation on the hands of the examiner when they move the passive joint to the supposed end of ROM. Active ROM is reported to have higher reliability than passive ROM, as passive measures depend on the force applied to the limbs by

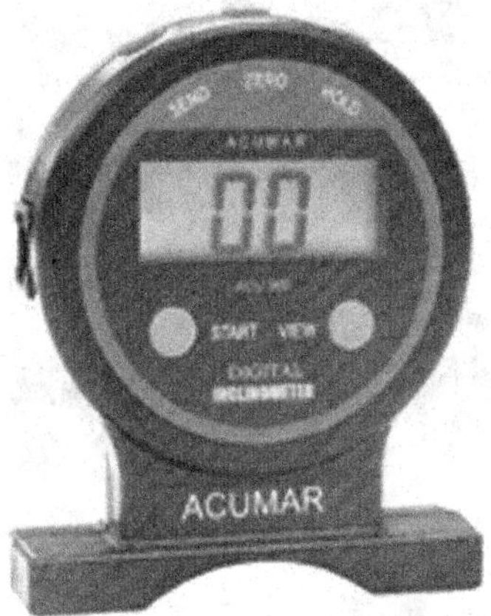

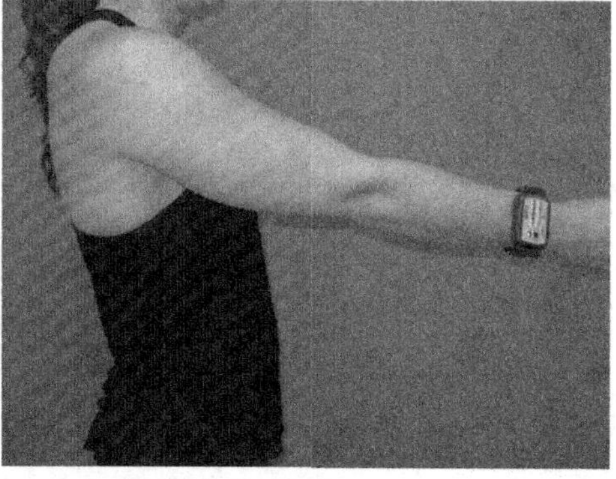

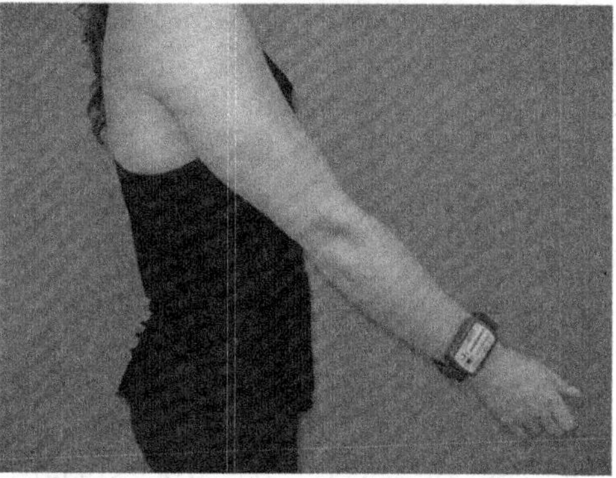

Figure 3.6 Inclinometers

another individual and that force and sensations can change unconsciously between trials and significantly between individuals.

Sometimes ROM tests may not measure what most people think they should measure. The ubiquitous sit and reach test is commonly used to measure lower back and hamstrings flexibility (see Figure 3.7). While a number of studies report that the sit and reach has moderate (95, 125) or strong (94) validity for measuring hamstrings extensibility but low validity for lumbar spine extensibility (117), another study contradicts and states that sit and reach (as well as the toe touch test) is a more appropriate measure of lumbar spinal flexibility and pelvic tilt ROM but not appropriate for hamstrings flexibility (144). These results instil confusion among the general population. The sit and reach test, and toe touch test can conceivably place undue pressure upon the vertebral discs (39) leading to injury. These concerns led to the development of the "back saver" sit and reach test, where only one leg (hamstring) is stretched at a time (157). Another variation is the chair sit and reach test (reliability: $r = 0.76–0.81$), where the individual while seated on the

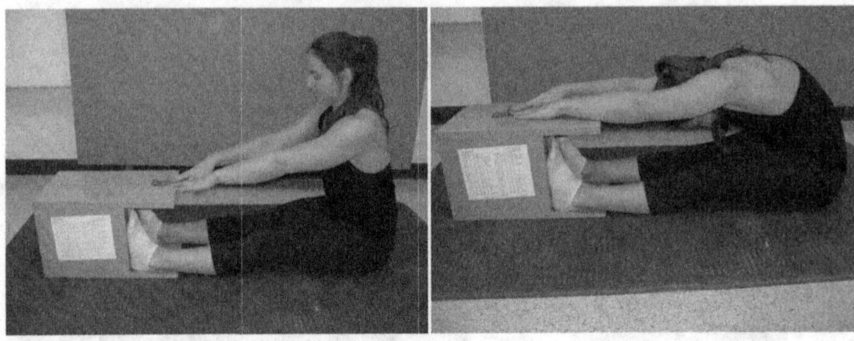

Figure 3.7 Sit and reach test

front edge of the chair extends one leg with other leg flexed to the side (97). The individual similarly reaches as far forward as possible. Furthermore, scapular abduction or arm length differences could also affect sit and reach scores (90). Hence, a "modified" sit and reach (91) was introduced where the score is based on the difference between the initial starting touch point of the fingers on the device from a straight back (good posture) seated position and their maximum reach rather than just the maximum reach, which is commonly measured in the non-modified sit and reach tests. Finally, it has also been argued that taller individuals can reach farther than shorter individuals with the sit and reach test, as a longer spine permits greater spinal flexion (67). Obviously, no test is perfectly valid or reliable but, in general, the articles tend to indicate that if you use one particular test throughout, then it will be fairly reliable at detecting change over time, but it is difficult and precarious to compare the scores between one test and another.

It is very important when monitoring flexibility to ensure that the joint or joints to be measured are isolated and that other joints are not contributing to a functional ROM. Individuals who lack flexibility or ROM in one or a series of joints sometimes can compensate by using or emphasizing another part of their body that has better flexibility. This compensatory relative flexibility (171) assumes that the individual will want to move through a movement or ROM with the least resistance. Thus, when measuring hip extensor ROM in a pronated position, if the individual has tight quadriceps, they may extend their lumbar spine to achieve a greater score or if measuring supine hip flexor ROM, with tight hamstrings they could posteriorly tilt their pelvis to help increase the movement of their leg. When performing an activity, the individual may compensate for their lack of flexibility by relying on another body segment not well designed for the movement and could lead to injuries (6).

Sex Differences

It is common knowledge that most women have better flexibility than the average man (4, 64, 79, 98, 183, 196). Some factors contributing to this difference could be differences in muscle mass, joint geometry and the degree of collagen in the musculotendinous unit (6). For instance, men with highly hypertrophied biceps brachii or hamstrings may be restricted by the muscle mass from achieving a full elbow flexion or knee flexion ROM. Adolescent female volleyball players demonstrated greater shoulder internal rotation ROM than men but there was no association of shoulder injuries with more restricted shoulder ROM (135). Women show greater hip flexibility with a single leg raise test (131) and sit and reach tests (115). The broader and shallower hips of women contribute to this greater flexibility (6). Pelvic and thoracic angles are also greater in women (118). Not all tests ratify these assumptions. For example, in one study, men demonstrated equal sit and reach test scores as women but the women had 8 percent greater pelvic flexion (118). Men tend to possess higher musculotendinous stiffness (89, 140), which would increase the resistance to a higher ROM. One study reported 44 percent greater gastrocnemius stiffness in men (139). Lower female passive muscle stiffness may be attributed to lesser female muscle cross-sectional area and thickness (111, 123) or an intrinsically more compliant female muscle (lower viscoelastic properties) (139). Female tendons also have greater compliance than men (110) with the greater female tendon compliance associated with their oestrogen secretion (35). Women are reported to have lower collagen fibril concentrations and percent area compared to males impacting the elasticity modulus (83). Collagen is a major component of skin, fascia, cartilage, ligaments, skin, tendons and bones. It is a protein composed of a triple helix, giving it high tensile strength. Reported decreased knee joint stiffness in women was attributed to differences in hormonal concentrations (e.g., oestrogen, progesterone) affecting ligamentous properties (more compliant as a result of less collagen) and associated with a greater Q angle (40). The Q angle is defined by drawing a line from the centre of the patella to the anterior superior iliac spine and another line from the tibial tubercle through the centre of the patella. The normal female value for this angle is 13–18 degrees and 12–15 degrees for males. Greater than normal Q angles could lead to chondromalacia patella, increased incidence of knee injuries or other symptoms.

Although sex differences in hormonal concentrations have been attributed to sex differences in joint flexibility, the literature is not unanimous regarding their effects on stretching. A recent study showed that significant acute increases in passive and active knee extension following static stretching (three x 45 seconds) were not influenced by the phases of the menstrual cycle (129). However, a Japanese study found that knee extension ROM and passive torque at the onset of pain were significantly increased during the ovulatory and luteal phases compared with the follicular phase. Passive stiffness decreased significantly during the ovulatory phase compared with the follicular phase (134). They reported that increases in muscle strength and activation (EMG) during the luteal phase was not related to flexibility, which suggested a greater implication for neural mechanisms.

Other endocrine differences occur with pregnancy. Women become more flexible during pregnancy owing to the release of various hormones such as relaxin, which allows greater extensibility of the interpubic ligament, however not all studies agree (6). Without this increased flexibility, the relatively large skull of the baby would never make it through the vaginal canal. This relaxin-induced increase in flexibility can also affect the ROM of other joints throughout the body (6).

After an acute bout of passive stretching (135 seconds), women demonstrated greater ROM increases, which were attributed to a greater stretch tolerance, as their musculotendinous stiffness did not change with the stretching (89). A study from Indonesia detected that static stretching was more effective for women whereas dynamic stretching was more favourable for men (115). Generally, in order to obtain similar ROM increases as women, men may need to stretch at a higher intensity or longer duration (89).

Ageing Differences

Ageing or senescence has been described as a "process of unfavourable progressive change" (112). Some older adults as they look in the mirror might counter that it is a depressive (psychologically and physiologically) rather than progressive change that alters their previously, younger, wrinkle free skin, decreases their strength, power and speed and increases their joint, ligament and musculotendinous stiffness. Older adults tend to be less flexible than their younger counterparts (60, 70). However, when they participate in stretch training programmes, their relative increases in flexibility are similar to young adults (60). Furthermore, trained older adults demonstrate greater degrees of flexibility than untrained older adults (60). As with other physical fitness parameters, the typically more sedentary lifestyle of senior adults exacerbates the decline in flexibility. A lack of activity can affect the synthesis, degradation and interconnections between connective tissue.

Collagen are long, fibrous structural proteins with high tensile strength and is the main component of fascia, cartilage, ligaments, tendons, bone and skin (36) (see Figure 3.8). It is an evolutionary ancient protein involved with the binding of cells of the simplest animals, such as sponges, as well as humans (11). Collagen has an extremely low compliance, similar to the tensile characteristics of copper. Intermolecular cross-links stabilize collagen, preventing the long rod-like molecules from sliding past each other forming nearly inextensible fibres (12). In combination with the proteins, elastin and soft keratin, the mix of these proteins provides not only strength (collagen) but also

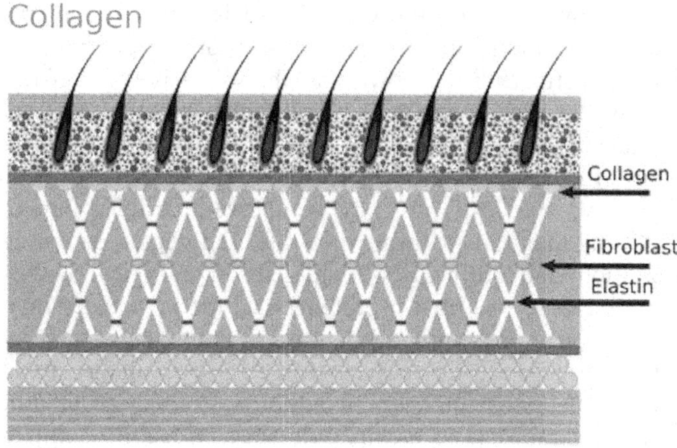

Figure 3.8 Collagen and elastin fibres within a tissue

elasticity (elastin) (5). Unlike collagen, elastin can double its length (12). With ageing, collagen increases markedly (5), which would have significant effects on ROM. The low compliance of collagen increases muscle, tendon and ligament stiffness. Thus, with increased stiffness or lower compliance, higher passive tension occurs for smaller increases in musculotendinous length. These changes would also have significant effects upon the stretch-shortening cycle. An increase in connective tissue proteins like collagen impede the muscle contraction/relaxation process, as it would have less extensible and compressible spring-like capabilities (5). A further complication is the excessive formation of intermolecular cross-links (11). Collagen and elastin cross-links in younger people promote strength and elasticity to the tissue but excessive cross-links with ageing can ensure that stiffness predominates over compliance. Another age-related change is the decreased hydration (water content) of aged tissue. Proteoglycans (i.e., chondroitin sulphate and keratin), which are present in virtually all extracellular matrices of connective tissues can hold large amounts of water, and so changes in their composition could lead to mild dehydration and some loss of function and extensibility (11). An example of this lack of extensibility with age can be seen with a simple test. Have a young person (especially a child) and a senior adult pinch the skin on the back of the hand. Then, quickly release the skin and be aware of the time it takes for the skin to return to its original shape and position. In a young child, it is practically instantaneous and would be almost impossible to measure with a stop watch or timer. With a senior adult, you can easily see and measure the slow return of the pinched skin to its original position. The older person's more dehydrated, cross linked, collagen predominant skin lacks the elasticity of the young person's skin. The same processes occur in the older person's connective tissues subcutaneously (under the skin) with the muscles, tendons, ligaments, and other tissues.

However as mentioned, this decreased ROM is attenuated in trained and more active older adults and with stretch training programmes. Coincidentally, animal studies have shown that the soleus muscles of rats do not get stiffer or as stiff with ageing and possess relatively, the same stiffness as young rat soleus muscles (5). How can this be? Rats in laboratories love to run on their little treadmills. The soleus muscle would be one of their most highly active muscles. Thus, just like active or trained humans, if the muscle activity is maintained throughout life, the degree of musculotendinous degradation and stiffness is decreased. In a typical inactive older adult, the more elastic-like elastin proteins degrade at a faster rate and are replaced by collagen. In senior aged, trained individuals, there would be less degradation, more elastin protein synthesis, less collagen replacement, and less unsuitable cross linkages. In addition, trained older adults have greater stretch tolerance (i.e., greater tolerance of passive joint moment), so they can push themselves farther while tolerating the relative discomfort (31). For older adults to remain flexible, it is recommended to participate in flexibility exercises at least two days each week for at least ten minutes each day (72). As you will read in the resistance training chapter, participating in resistance training even without stretching can enhance ROM (3). Whether older adults use traditional resistance training or functional training (multi-articular and multiplanar exercises to improve movement ability, central body strength and neuromuscular efficiency), both provided similar improvements in joint mobility over a 12-week programme (49). An American training programme called the Strong Women Programme (progressive resistance training, balance training, and flexibility exercises) showed significant sit and reach flexibility increases over their twice weekly, 12-week programme (175).

Youth Differences

It is generally considered that children have better flexibility than adults. However, more age-specific research has shown different patterns with prepubescence and pubescence. Younger children around the age of five years typically show a high level of flexibility which declines progressively until the age of 12 years. During puberty until pre-adulthood (12–18 years), their flexibility exhibits improvements (38, 76, 99, 105). Even with very young children when comparing 5-year-old versus 6-year-old children (64) or kindergarten to second grade (132) there has been evidence of decreases in flexibility. Not all studies provide a consistent picture. In contrast to the early studies that reported improved flexibility throughout puberty, others have reported decrements (45, 55, 68, 108, 138, 173) or no significant change (37, 128). Similar to senior adults who experience decrements in strength, power and ROM, much of the impairments can be related to their activity levels. A similar relationship seems to exist with children with the less active experiencing great ROM deficits. Young and pubescent children who partake in extensive flexibility programmes such as gymnasts, dancers, figure skaters and others can exhibit astounding levels of flexibility, continually improving from young childhood and through puberty. Other, less active children may lose their natural flexibility. owing to the time spent sitting in school and at home (59, 133). Stretching programmes can start very early especially for extreme ROM sports like gymnastics. A flexibility programme for 5–6-year-old female gymnasts called "Game stretching", which involves the use of balls, hoops, ribbons, maces, and other items while pretending to imitate different animals, flowers or actions (e.g., fish, dolphin, starfish, mouse, tiger, "frog", butterfly, ballerina, cobra, snail, swing, swimmer, ostrich, fox, boa, and others) was effective in improving the ROM of various joints (51). Youth classical dancers (~nine years old) improved their passive hip flexion ROM similarly when implementing either static or PNF stretching (147).

It has been suggested that diminished flexibility during puberty as well as so-called "growth pains", and "tightness" might be attributed to a greater growth rate of the skeletal bones compared to the growth rate of muscles and connective tissue (9, 92), (113, 130). However, a number of studies dispute this common belief citing in one study that older adolescents actually had greater flexibility (163) and a study of 600 13–14-year-old students indicated that whereas a decrease in flexibility is associated with growth, growth does not reduce flexibility (59). Furthermore, "growth pains" normally occur prior to the peak height velocity stage and thus have no connection with growth (145). Regardless of the association or lack of association with growth, one study reported that stretching reduced the duration of "growing pains" in 5–14-year-old children (18).

Genetic Differences

Claude Bouchard originally from Quebec, Canada is internationally renowned for his early genetic studies with twins. His review (34) examined the effect of genetics on flexibility. The flexibility relationship in his studies and others were quite low. Whereas two studies from the same researcher reported moderate to strong correlations between 11–15-year-old male twins (0.69) (106) and 12–17-year-old male and female twins for trunk, hip and shoulder flexibility (0.7–0.91) (106), most other twin studies found weak correlations (0.18–0.43) (42, 53, 159, 160). These results again point to the importance of the environment (nature vs. nurture); that is, the activity levels and flexibility training of individuals plays a more important role than their genetics for affecting ROM.

Limb Dominance Differences

A similar activity-related rationale may be attributed to reports of differences in ROM with dominant versus non-dominant limbs. Two tennis studies (41, 44) reported decreased internal rotation of the shoulder but increased external rotation ROM of the dominant arm. Two baseball studies reported either no major differences between right and left sides (77) versus greater hip flexion and internal rotation of the stance leg with baseball pitchers (188). There is conflict in the literature with some studies showing insignificant or small side ROM differences with pubescent and young adult females' shoulders and lower extremities (105) and no differences in ankle dorsiflexion and plantar flexion ROM with 15–34-year-olds (142). Other studies report lateralized differences with less right wrist mobility and left hip rotation (4), and greater dominant limb humeral head retroversion angle (109). Rather than limb side differences in ROM being due to a lateralized predisposition, it seems that differences are more likely attributed to a specific, unilateral, expanded dynamic ROM (increased ROM) or higher incidence of injury to a predominant limb during repetitive tasks (decreased ROM).

Circadian (Diurnal: time of day) Differences

Restricted ROM seems to be more prevalent in the morning (17, 32, 54, 69, 150, 152, 153, 166, 194). As the individual becomes more active during the day and increases their core temperature, there will be decreases in visco-elasticity (thixotropic effects: see next section for more details), increased tissue compliance and less resistance to motion. Diurnal fluctuations in endocrine responses (i.e., epinephrine, norepinephrine, thyroid hormones, testosterone, insulin-like growth factors, growth hormone) (107) would also impact basal metabolic activity contributing to core heat flux changes during the day. As the individual becomes less active and the endocrine activity subsides later in the evening in preparation for sleeping, flexibility also diminishes (69). In order to maintain, a suitable ROM towards the evening, an individual would need to maintain or increase activity in an attempt to maintain a higher body temperature and sustain the energy facilitating hormones (i.e., epinephrine, norepinephrine, thyroid hormones, testosterone, insulin-like growth factors, growth hormone and others).

Summary

Flexibility, defined as the ROM around a joint can be altered with passive and active static, dynamic and PNF stretching as well as other techniques. These stretching techniques can provide elastic (acute or non-permanent) or plastic (semi-permanent with prolonged training) changes to joint flexibility. The majority of studies show that static and PNF stretching provide greater increases in ROM versus dynamic stretching. Yoga is also very effective for improving ROM, although yoga integrates more diverse activities than just stretching (i.e., breathing techniques, static and prolonged static contractions and meditation among other activities), which can affect overall relaxation (enhanced parasympathetic stimulation). Normative data for joint ROM are provided in a number of texts and articles. Most measurement techniques such as the use of goniometers, fleximeters, inclinometers, photography, sit (stand) and reach tests display high reliability but may not provide similar between device values.

Females tend to have greater joint ROM than men, owing to anatomical differences, less musculotendinous unit stiffness (greater compliance), and endocrine differences. With ageing, people tend to become less flexible which may be related to increased collagen proportions and protein cross-linkages. However, much of the flexibility impairments can be attributed to greater inactivity. Youth, for their part, tend to have higher flexibility levels which may decrease during puberty compared to childhood. Once again, though, the lower levels of flexibility with puberty can be related to more inactivity. Regarding the effect of genetics on flexibility, activity again plays a more important role than DNA as studies have shown weak to strong correlations. Individuals tend to be less flexible in the morning, as core temperature is lower and the relatively long period of inactivity tends to restrict movement.

References

1. Al Attar, W.S.A., Bizzini, M., Alkabkabi, F., Alshamrani, N., Alarifi, S., Alzahrani, H., Ghulam, H., Aljedaani, E., and Sanders, R.H.Effectiveness of the FIFA 11+ Referees Injury Prevention Program in reducing injury rates in male amateur soccer referees. *Scand J Med Sci Sports* 31: 1774–1781, 2021.
2. Al Attar, W.S.A., Faude, O., Bizzini, M., Alarifi, S., Alzahrani, H., Almalki, R.S., Banjar, R.G., and Sanders, R.H.The FIFA 11+ Shoulder Injury Prevention Program Was Effective in Reducing Upper Extremity Injuries Among Soccer Goalkeepers: A Randomized Controlled Trial. *Am J Sports Med* 49: 2293–2300, 2021.
3. Alizadeh, S., Daneshjoo, A., Zahiri, A., Anvar, S.H., Goudini, R., Hicks, J.P., Konrad, A., and Behm, D.G.Resistance Training Induces Improvements in Range of Motion: A Systematic Review and Meta-Analysis. *Sports Med* 53: 707–722, 2023.
4. Allander, E., Bjornsson, O.J., Olafsson, O., Sigfusson, N., and Thorsteinsson, J.Normal range of joint movements in shoulder, hip, wrist and thumb with special reference to side: a comparison between two populations. *Int J Epidemiol* 3: 253–261, 1974.
5. Alnaqeeb, M.A., Al Zaid, N.S., and Goldspink, G.Connective tissue changes and physical properties of developing and ageing skeletal muscle. *J Anat* 139 (Pt 4): 677–689, 1984.
6. Alter, M.J. *Science of Flexibility*. Champaign. IL: Human Kinetics, 1996.
7. American College of Sports Medicine. ACSM's Guidelines for Exercise Testing and Prescription. *Medicine Science Sports and Exercise* 6: 158–164, 2000.
8. Amiri-Khorasani, M., Abu Osman, N.A., and Yusof, A.Acute effect of static and dynamic stretching on hip dynamic range of motion during instep kicking in professional soccer players. *Journal of Strength and Conditioning Research/National Strength & Conditioning Association* 25: 1647–1652, 2011.
9. Bachrach, R.M.Injuries to dancer's spine. In: *Dance medicine*, A.J. Ryan and R.E. Stephens (Eds). Chicago: Pluribus Press, 1987, pp. 243–266.

10. Bacurau, R.F., Monteiro, G.A., Ugrinowitsch, C., Tricoli, V., Cabral, L.F., and Aoki, M.S.Acute effect of a ballistic and a static stretching exercise bout on flexibility and maximal strength. *J Strength Cond Res* 23: 304–308, 2009.
11. Bailey, A.J.Molecular mechanisms of ageing in connective tissues. *Mech Ageing Dev* 122: 735–755, 2001.
12. Bailey, A.J., Light, N.D., and Atkins, E.D.Chemical cross-linking restrictions on models for the molecular organization of the collagen fibre. *Nature* 288: 408–410, 1980.
13. Balaji, P.A., Varne, S.R., and Ali, S.S.Physiological effects of yogic practices and transcendental meditation in health and disease. *N Am J Med Sci* 4: 442–448, 2012.
14. Bandy, W.D., Irion, J.M., and Briggler, M.The effect of time and frequency of static stretching on flexibility of the hamstring muscles. *Physical Therapy* 77: 1090–1096, 1997.
15. Bandy, W.D., Irion, J.M., and Briggler, M.The effect of static stretch and dynamic range of motion training on the flexibility of the hamstring muscles. *J Orthop Sports PhysTher* 27: 295–300, 1998.
16. Barroso, R., Tricoli, V., Santos Gil, S.D., Ugrinowitsch, C., and Roschel, H.Maximal strength, number of repetitions, and total volume are differently affected by static-, ballistic-, and proprioceptive neuromuscular facilitation stretching. *J Strength Cond Res* 26: 2432–2437, 2012.
17. Baxter, C. and Reilly, T.Influence of time of day on all-out swimming. *Br J Sports Med* 17: 122–127, 1983.
18. Baxter, M.P. and Dulberg, C."Growing pains" in childhood: A proposed treatment. *Journal of Pediatric Orthopaedics* 4: 402–406, 1988.
19. Beedle, B.B. and Mann, C.L.A comparison of two warm-ups on joint range of motion. *JStrength CondRes* 21: 776–779, 2007.
20. Behm, D.G., Alizadeh, S., Anvar, S.H., Drury, B., Granacher, U., and Moran, J. Non-local Acute Passive Stretching Effects on Range of Motion in Healthy Adults: A Systematic Review with Meta-analysis. *Sports Med* 51: 945–959, 2021.
21. Behm, D.G., Alizadeh, S., Daneshjoo, A., Anvar, S.H., Graham, A., Zahiri, A., Goudini, R., Edwards, C., Culleton, R., Scharf, C., and Konrad, A.Acute Effects of Various Stretching Techniques on Range of Motion: A Systematic Review with Meta-Analysis. *Sports Med Open* 9: 107, 2023.
22. Behm, D.G., Alizadeh, S., Drury, B., Granacher, U., and Moran, J.Non-local acute stretching effects on strength performance in healthy young adults. *Eur J Appl Physiol* 121: 1517–1529, 2021.
23. Behm, D.G., Bambury, A., Cahill, F., and Power, K.Effect of acute static stretching on force, balance, reaction time, and movement time. *Med Sci Sports Exerc* 36: 1397–1402, 2004.
24. Behm, D.G., Blazevich, A.J., Kay, A.D., and McHugh, M.Acute effects of muscle stretching on physical performance, range of motion, and injury incidence in healthy active individuals: a systematic review. *Appl Physiol Nutr Metab* 41: 1–11, 2016.
25. Behm, D.G., Button, D.C., and Butt, J.C.Factors affecting force loss with prolonged stretching. *Can J Appl Physiol* 26: 261–272, 2001.
26. Behm, D.G. and Chaouachi, A.A review of the acute effects of static and dynamic stretching on performance. *Eur J Appl Physiol* 111: 2633–2651, 2011.
27. Behm, D.G., Plewe, S., Grage, P., Rabbani, A., Beigi, H.T., Byrne, J.M., and Button, D.C.Relative static stretch-induced impairments and dynamic stretch-induced enhancements are similar in young and middle-aged men. *Appl Physiol Nutr Metab* 36: 790–797, 2011.

28. Behm, D.G. and Sale, D.G.Velocity specificity of resistance training. *Sports Med* 15: 374–388, 1993.
29. Behm, D.G., Kay, A.D., Trajano, G.S., Alizadeh, S., and Blazevich, A.J.Effects of stretching on injury risk reduction and balance. *Journal of Clinical Exercise Physiology* 10: 106–116, 2021.
30. Bishop, D.Warm up II: performance changes following active warm up and how to structure the warm up. *Sports Med – ADIS Int* 33: 483–498, 2003.
31. Blazevich, A.J., Cannavan, D., Waugh, C.M., Fath, F., Miller, S.C., and Kay, A.D. Neuromuscular factors influencing the maximum stretch limit of the human plantar flexors. *Journal of Applied Physiology* 113: 1446–1455, 2012.
32. Bompa, T. *Theory and methodology of training*. Dubuque, IA: Kendall/Hunt, 1990.
33. Boone, D.C. and Azen, S.P.Normal range of motion of joints in male subjects. *J Bone Joint Surg Am* 61: 756–759, 1979.
34. Bouchard C., Malina, R.M., and Perusse, L. *Genetics of fitness and physical performance* Champaign, IL: Human Kinetics, 1997.
35. Bryant, A.L., Clark, R.A., Bartold, S., Murphy, A., Bennell, K.L., Hohmann, E., Marshall-Gradisnik, S., Payne, C., and Crossley, K.M.Effects of estrogen on the mechanical behavior of the human Achilles tendon in vivo. *J Appl Physiol (1985)* 105: 1035–1043, 2008.
36. Buehler, M.J.Nature designs tough collagen: explaining the nanostructure of collagen fibrils. *Proc Natl Acad Sci U S A* 103: 12285–12290, 2006.
37. Burley L.R., Dobell, H.C., and Farrell, B.J. Relations of power, speed, flexibility, and certain anthropometric measures of junior high school girls. *Research Quarterly* 32: 442–448, 1961.
38. Buxton, D.Extension of the Kraus-Weber test. *Research Quarterly* 3: 210–217, 1957.
39. Cailliet, R. *Low back pain syndrome* Philadelphia, PA: F.A. Davis, 1988.
40. Cammarata, M.L. and Dhaher, Y.Y.The differential effects of gender, anthropometry, and prior hormonal state on frontal plane knee joint stiffness. *Clin Biomech (Bristol, Avon)* 23: 937–945, 2008.
41. Chandler, T.J., Kibler, W.B., Uhl, T.L., Wooten, B., Kiser, A., and Stone, E.Flexibility comparisons of junior elite tennis players to other athletes. *Am J Sports Med* 18: 134–136, 1990.
42. Chatterjee, S. and Das, N.Physical and motor fitness in twins. *Jpn J Physiol* 45: 519–534, 1995.
43. Chaves, T.C, Nagamine, H.M., Belli, J.F.C., de Hannai, M.C.T., Bevilacqua-Grossi, D., and de Oleivera, O.S.Reliability of fleximetry and goniometry for assessing cervical range of motion among children. *Review Brazilian Fisioterapa* 12: 183–190, 2008.
44. Chinn C.J., Priest J.D., and Kent B.E.Upper extremity range of motion, grip strength, and girth in highly skilled tennis players. *Phys Ther* 54: 474–483, 1974.
45. Clarke, H.H.Joint and body range of movement. *Physical Fitness Research Digest* 5: 16–18, 1975.
46. Condon, S.M. and Hutton, R.S.Soleus muscle electromyographic activity and ankle dorsiflexion range of motion during four stretching procedures. *Physical Therapy* 67: 24–30, 1987.
47. Covert, C.A., Alexander, M.P., Petronis, J.J., and Davis, D.S.Comparison of Ballistic and Static Stretching on Hamstring Muscle Length Using an Equal Stretching Dose. *JStrength CondRes* 24: 3008–3014, 2010.

48. Cronin, J., Nash, M., and Whatman, C.The acute effects of hamstring stretching and vibration on dynamic knee joint range of motion and jump performance. *Phys Ther Sport* 9: 89–96, 2008.
49. de Resende-Neto, A.G., do Nascimento, M.A., de Sa, C.A., Ribeiro, A.S., Desantana, J.M., and da Silva-Grigoletto, M.E.Comparison between functional and traditional training exercises on joint mobility, determinants of walking and muscle strength in older women. *J Sports Med Phys Fitness* 59: 1659–1668, 2019.
50. de Vries, H.A. *Physiology of exercise.* Duboque, IA: Wm. C. Brown, 1986.
51. Deineko, A., Prusik, K., Krasova, I., Batieieva, N., and Marchenkov, M.Game stretching as a modern means of developing the flexibility of 5–6-year-old female gymnasts. *Slobozhanskyi Herald of Science and Sport* 26: 88–94, 2022.
52. Depino, G.M., Webright, W.G., and Arnold, B.L.Duration of maintained hamstring flexibility after cessation of an acute static stretching protocol. *J Athl Train* 35: 56–59, 2000.
53. Devor, E.J. and Crawford, M.H.Family resemblance for neuromuscular performance in a Kansas Mennonite community. *Am J Phys Anthropol* 64: 289–296, 1984.
54. Dick, F.W. *Sports training principles.* London: Lepus Bokks, 1980.
55. Docherty, D. and Bell, R.D.The relationship between flexibility and linearity measures in boys and girls 6–15 years of age. *Journal of Human Movement Studies* 11: 279–288, 1985.
56. Duncan, M.J. and Woodfield, L.A.Acute effects of warm-up protocol on flexibility and vertical jump in children. *Journal of Exercise Physiology* 9: 9–16, 2006.
57. Etnyre, B.R. and Lee, E.J.Chronic and Acute Flexibility of Men and Women Using 3 Different Stretching Techniques. *Research Quarterly for Exercise and Sport* 59: 222–228, 1988.
58. Fasen, J.M., O'Connor, A.M., Schwartz, S.L., Watson, J.O., Plastaras, C.T., Garvan, C.W., Bulcao, C., Johnson, S.C., and Akuthota, V.A randomized controlled trial of hamstring stretching: comparison of four techniques. *J Strength Cond Res* 23: 660–667, 2009.
59. Feldman, D., Shrier, I., Rossignol, M., and Abenhaim, L.Adolescent growth is not associated with changes in flexibility. *Clin J Sport Med* 9: 24–29, 1999.
60. Ferber, R., Osternig, L., and Gravelle, D.Effect of PNF stretch techniques on knee flexor muscle EMG activity in older adults. *J Electromyogr Kinesiol* 12: 391–397, 2002.
61. Fletcher, I.M.The effect of different dynamic stretch velocities on jump performance. *Eur J Appl Physiol* 109: 491–498, 2010.
62. Fletcher, I.M. and Jones, B.The effect of different warm-up stretch protocols on 20 meter sprint performance in trained rugby union players. *J Strength Cond Res* 18: 885–888, 2004.
63. Fowles, J.R., Sale, D.G., and MacDougall, J.D.Reduced strength after passive stretch of the human plantar flexors. *Journal of Applied Physiology* 89: 1179–1188, 2000.
64. Gabbard, C.T. and Tandy, R.Body composition and flexibility amoung prepubescent males and females. *Journal of Human Movement Studies* 4: 153–159, 1988.
65. Gajdosik, R.L. and Bohannon, R.W.Clinical measurement of range of motion. Review of goniometry emphasizing reliability and validity. *Phys Ther* 67: 1867–1872, 1987.
66. Gajdosik, R.L., Vander Linden, D.W., and Williams, A.K.Influence of age on length and passive elastic stiffness characteristics of the calf muscle-tendon unit of women. *Phys Ther* 79: 827–838, 1999.

67. Gatton, M.L. and Pearcy, M.J.Kinematics and movement sequencing during flexion of the lumbar spine. *Clin Biomech (Bristol, Avon)* 14: 376–383, 1999.
68. Germain, N.W. and Blair, S.N.Variability of shoulder flexion with age, activity and sex. *Am Correct Ther J* 37: 156–160, 1983.
69. Gifford, L.S.Circadian variation in human flexibility and grip strength. *Aust J Physiother* 33: 3–9, 1987.
70. Gonzalez-Rave, J.M., Sanchez-Gomez, A., and Santos-Garcia, D.J.Efficacy of two different stretch training programs (passive vs. proprioceptive neuromuscular facilitation) on shoulder and hip range of motion in older people. *J Strength Cond Res* 26: 1045–1051, 2012.
71. Goodwin, J., Clark, C., Deakes, J., Burdon, D., and Lawrence, C.Clinical methods of goniometry: a comparative study. *Disabil Rehabil* 14: 10–15, 1992.
72. Granacher, U. and Hortobagyi, T.Exercise to Improve Mobility in Healthy Aging. *Sports Med* 45: 1625–1626, 2015.
73. Guissard, N. and Duchateau, J.Effect of static stretch training on neural and mechanical properties of the human plantar-flexor muscles. *Muscle and Nerve* 29: 248–255, 2004.
74. Guissard, N. and Duchateau, J.Neural aspects of muscle stretching. *ExercSport SciRev* 34: 154–158, 2006.
75. Guissard, N., Duchateau, J., and Hainaut, K.Muscle stretching and motoneuron excitability. *European Journal of Applied Physiology* 58: 47–52, 1988.
76. Gurewitsch, A.D. and O'Neill, M.Flexibility of healthy children. *Archives of Physical Therapy* 4: 216–221, 1994.
77. Gurry, M., Pappas, A., Michaels, J., Maher, P., Shakman, A., Goldberg, R., and Rippe, J.A Comprehensive Preseason Fitness Evaluation for Professional Baseball Players. *Phys Sportsmed* 13: 63–74, 1985.
78. Halbertsma, J., Bolhuis, A., and Goeken, L.Sport stretching: effect of passive muscle stiffness of short hamstrings. *Archives of Physical Medicine and Rehabilitation* 77: 688–692, 1996.
79. Haley, S.M., Tada, W.L., and Carmichael, E.M.Spinal mobility in young children. A normative study. *Phys Ther* 66: 1697–1703, 1986.
80. Hallaceli, H., Uruc, V., Uysal, H.H., Ozden, R., Hallaceli, C., Soyuer, F., Ince Parpucu, T., Yengil, E., and Cavlak, U.Normal hip, knee and ankle range of motion in the Turkish population. *Acta Orthop Traumatol Turc* 48: 37–42, 2014.
81. Halvorson, G.A.Principles of rehabilitating sports injuries. In: *Scientific foundations of sports medicine*, C.C. Teitz (Ed.). Philadelphia, PA: Decker, 1989, pp. 345–371.
82. Hammes, D., Aus der Fünten, K., Kaiser, S., Frisen, E., Bizzini, M., and Meyer, T. Injury prevention in male veteran football players – a randomised controlled trial using "c". *J Sports Sci* 33: 873–881, 2015.
83. Hashemi, J., Chandrashekar, N., Mansouri, H., Slauterbeck, J.R., and Hardy, D. M.The human anterior cruciate ligament: sex differences in ultrastructure and correlation with biomechanical properties. *J Orthop Res* 26: 945–950, 2008.
84. Hayes, K., Walton, J.R., Szomor, Z.R., and Murrell, G.A.Reliability of five methods for assessing shoulder range of motion. *Aust J Physiother* 47: 289–294, 2001.
85. Hebbelink, M. *Flexibility*. Oxford: Blackwell Scientific, 1988.
86. Heyward, V.H. *Advanced Fitness Assessment and Exercise Prescription*. Windsor, ON: Human Kinetics Publ., 2005.

87. Hindle, K.B, Whitcomb, T.J., Briggs, W.O., and Hong, J.Proprioceptive neuromuscular facilitation (PNF): Its mechanisms and effects on range of motion and muscular function. *Journal of Human Kinetics* 31: 105–113, 2012.
88. Hoffman, J. *Norms for Fitness, Performance, and Health.* Champaign, IL: Human Kinetics Publishers, 2006.
89. Hoge, K.M., Ryan, E.D., Costa, P.B., Herda, T.J., Walter, A.A., Stout, J.R., and Cramer, J.T.Gender differences in musculotendinous stiffness and range of motion after an acute bout of stretching. *J Strength Cond Res* 24: 2618–2626, 2010.
90. Hopkins, D.R.The relationship between selected anthropometric measures and sit-and-reach performance. In: *Dance National Measurement Symposium*. Houston, TX, 1981.
91. Hopkins, D.R. and Hoeger, W.W.The modified sit and reach test. In: *Lifetime physical fitness and wellness: a personalized program.* Englewood, CO: Morton, 1986, pp. 47–48.
92. Howse, A.J.G.The young ballet dancer. In: *Dance medicine: A comprehensive guide*, A.J. Ryan and R.E. Stephens (Ed.). Chicago, IL: Pluribus Press, 1987, pp. 107–114.
93. Hubley-Kozey, C.L.Testing Flexibility. In: *Physiological Testing of the High Performance Athlete*, J.D. MacDougall, H.A. Weger, and H.J. Green (Eds). Champaign, IL: Human Kinetics, 1991, pp. 309–359.
94. Jackson, A. and Langford, N.J.The criterion-related validity of the sit and reach test: replication and extension of previous findings. *Res Q Exerc Sport* 60: 384–387, 1989.
95. Jackson, A.W. and Baker, A.A.The relationship of the sit and reach test to criterion measures of hamstring and back flexibility in young females. *Research Quarterly for Exercise and Sport* 3: 183–186, 1986.
96. Jensen, A.M., Ramasamy, A., and Hall, M.W.Improving general flexibility with a mind-body approach: a randomized, controlled trial using neuro emotional Technique(R). *J Strength Cond Res* 26: 2103–2112, 2012.
97. Jones, C.J., Rikli, R.E., Max, J., and Noffal, G.The reliability and validity of a chair sit-and-reach test as a measure of hamstring flexibility in older adults. *Res Q Exerc Sport* 69: 338–343, 1998.
98. Jones, M.A., Buis, J.M., and Harris, I.D.Relationship of race and sex to physical and motor measures. *Perceptual and Motor Skills* 1: 169–170, 1986.
99. Kendall, H.O. and Kendall, F.P.Normal flexibility according to age groups. *J Bone Joint Surg Am* 30A: 690–694, 1948.
100. Kent, M. *The Oxford dictionary of sports science and medicine*. Oxford: Oxford University Press, 1998.
101. Kisner, C. and Colby, L.A. *Therapeutic exercise foundations and techniques.* Philadelphia, PA: F.A. Davis, 2002.
102. Knudson, D.Stretching during warm-up: do we have enough evidence? *JOPERD* 70: 24–28, 1999.
103. Konrad, A., Alizadeh, S., Daneshjoo, A., Anvar, S.H., Graham, A., Zahiri, A., Goudini, R., Edwards, C., Scharf, C., and Behm, D.G.Chronic effects of stretching on range of motion with consideration of potential moderating variables: A systematic review with meta-analysis. *J Sport Health Sci*: 2023.
104. Konrad, A., Stafilidis, S., and Tilp, M.Effects of acute static, ballistic, and PNF stretching exercise on the muscle and tendon tissue properties. *Scand J Med Sci Sports* 27: 1070–1080, 2017.
105. Koslow, R.E.Bilateral flexibility in the upper and lower extremities. *Journal of Human Movement Studies* 9: 467–472, 1987.

106. Kovar, R. *Human variation in motor abilities and genetic analysis.* Prague: Charles University, 1981.
107. Kraemer, W.J., Ratamess, N.A., and Nindl, B.C.Recovery responses of testosterone, growth hormone, and IGF-1 after resistance exercise. *J Appl Physiol (1985)* 122: 549–558, 2017.
108. Krahenbuhl, G.S. and Martin, S.L.Adolescent body size and flexibility. *Res Q* 48: 797–799, 1977.
109. Kronberg, M., Brostrom, L.A., and Soderlund, V.Retroversion of the humeral head in the normal shoulder and its relationship to the normal range of motion. *Clin Orthop Relat Res*: 113–117, 1990.
110. Kubo, K., Kanehisa, H., and Fukunaga, T.Gender differences in the viscoelastic properties of tendon structures. *European Journal of Applied Physiology and Occupational Physiology* 88: 520–526, 2003.
111. Kubo, K., Kanehisa, H., Ito, M., and Fukunaga, T.Effects of isometric training on the elasticity of human tendon structures in vivo. *J Appl Physiol (1985)* 91: 26–32, 2001.
112. Lansing, A.I.Some physiological aspects of ageing. *Physiological Reviews* 31: 274–278, 1978.
113. Leard, J.S.Flexibility and conditioning in the young athlete. In: *Pediatric and adolescent sport medicine*, L.J. Micheli (Ed.). Boston, MA: Little, Brown, 1984, pp. 194–210.
114. Leighton, B. *Manual of Instruction for the Leighton Flexometer*, 1987.
115. Lestari, A.Y., Abdul, A., Tomoliyus, A., Sukamti, E.R., and Hartanto, A.Static vs dynamic stretching: which is better for flexibility in terms of gender of badminton athletes? *Pedagogy of Physical Culture and Sports* 27: 368–377, 2023.
116. Liemohn, W.P.Flexibility and Muscular Strength. *Journal of Physical Education, Recreation and Dance* 59: 37–40, 1988.
117. Liemohn, W.P., Sharpe, G.L., and Wasserman, J.F.Lumbosacral movement in the sit-and-reach and in Cailliet's protective-hamstring stretch. *Spine (Phila Pa 1976)* 19: 2127–2130, 1994.
118. Lopez-Minarro, P.A., Andujar, P.S., and Rodriguez-Garcia, P.L.A comparison of the sit-and-reach test and the back-saver sit-and-reach test in university students. *J Sports Sci Med* 8: 116–122, 2009.
119. Madanmohan, Thombre, D.P., Balakumar, B., Nambinarayanan, T.K., Thakur, S., Krishnamurthy, N., and Chandrabose, A.Effect of yoga training on reaction time, respiratory endurance and muscle strength. *Indian J Physiol Pharmacol* 36: 229–233, 1992.
120. Maddigan, M.E., Peach, A.A., and Behm, D.G.A comparison of assisted and unassisted proprioceptive neuromuscular facilitation techniques and static stretching. *Journal of strength and conditioning research/National Strength & Conditioning Association* 26: 1238–1244, 2012.
121. Magnusson, M. and Pope, M.H.Body height changes with hyperextension. *Clinical Biomechanics* 4: 236–238, 1996.
122. Magnusson, S.P., Aagaard, P., and Nielson, J.J.Passive energy return after repeated stretches of the hamstring muscle-tendon unit. *Med Sci Sports Exerc* 32: 1160–1164, 2000.
123. Magnusson, S.P., Simonsen, E.B., Aagaard, P., Boesen, J., Johannsen, F., and Kjaer, M.Determinants of musculoskeletal flexibility: viscoelastic properties, cross-sectional area, EMG and stretch tolerance. *Scand J Med Sci Sports* 7: 195–202, 1997.

124. Marshall, P.W., Cashman, A., and Cheema, B.S.A randomized controlled trial for the effect of passive stretching on measures of hamstring extensibility, passive stiffness, strength, and stretch tolerance. *J Sci Med Sport* 14: 535–540, 2011.
125. Mayorga-Vega, D., Merino-Marban, R., and Viciana, J.Criterion-Related Validity of Sit-and-Reach Tests for Estimating Hamstring and Lumbar Extensibility: a Meta-Analysis. *J Sports Sci Med* 13: 1–14, 2014.
126. McHugh, M.P. and Cosgrave, C.H.To stretch or not to stretch: the role of stretching in injury prevention and performance. *Scand J Med Sci Sports* 20: 169–181, 2010.
127. Medeiros, D.M.M.D.Comparison between static stretching and proprioceptive neuromuscular facilitation on hamstring flexibility: Systematic review and meta-analysis. *European Journal of Physiotherapy*, 2018.
128. Mellin, G. and Poussa, M.Spinal mobility and posture in 8- to 16-year-old children. *J Orthop Res* 10: 211–216, 1992.
129. Michalik, P., Michalski, T., Witkowski, J., and Widuchowski, W.The influence of menstrual cycle on the efficiency of stretching. *Adv Clin Exp Med* 31: 381–387, 2022.
130. Micheli, L.J.Overuse injuries in children's sports: the growth factor. *Orthop Clin North Am* 14: 337–360, 1983.
131. Mier, C.M. and Shapiro, B.S.Sex differences in pelvic and hip flexibility in men and women matched for sit-and-reach score. *J Strength Cond Res* 27: 1031–1035, 2013.
132. Milne, C., Seefeldt, V., and Reuschlein, P.Relationship between grade, sex, race, and motor performance in young children. *Res Q* 47: 726–730, 1976.
133. Milne, R.A. and Mierau, D.R.Hamstrings distensibility in the general population: Relationship to pelvic and low back stresses. *Journal of Manipulative and Physiological Therapeutics* 2: 146–150, 1979.
134. Miyazaki, M. and Maeda, S.Changes in hamstring flexibility and muscle strength during the menstrual cycle in healthy young females. *The Journal of Physical Therapy Science* 34: 92–98, 2022.
135. Mizoguchi, Y., Suzuki, K., Shimada, N., Naka, H., Kimura, F., and Akasaka, K. Prevalence of Glenohumeral Internal Rotation Deficit and Sex Differences in Range of Motion of Adolescent Volleyball Players: A Case-Control Study. *Healthcare (Basel)* 10, 2022.
136. Mizuno, T., Matsumoto, M., and Umemura, Y.Decrements in stiffness are restored within 10 min. *Int J Sports Med* 34: 484–490, 2013.
137. Moller, M., Oberg, B., Ekstrand, J., and Gillquist, J.Effects of warming up, massage, and stretching on range of motion and muscle strength in the lower extremity. *The American Journal of Sports Medicine* 11: 249–252, 1983.
138. Moran, H.M., Hall, M.A., Barr, A., and Ansell, B.M.Spinal mobility in the adolescent. *Rheumatol Rehabil* 18: 181–185, 1979.
139. Morse, C.I.Gender differences in the passive stiffness of the human gastrocnemius muscle during stretch. *European Journal of Applied Physiology* 111: 2149–2154, 2011.
140. Morse, C.I.Gender differences in the passive stiffness of the human gastrocnemius muscle during stretch. *Eur J Appl Physiol* 111: 2149–2154, 2011.
141. Morton, S.K., Whitehead, J.R., Brinkert, R.H., and Caine, D.J.Resistance training vs. static stretching: effects on flexibility and strength. *J Strength Cond Res* 25: 3391–3398, 2011.
142. Moseley, A.M., Crosbie, J., and Adams, R.Normative data for passive ankle plantarflexion–dorsiflexion flexibility. *Clin Biomech (Bristol, Avon)* 16: 514–521, 2001.

143. Mullaney, M.J., McHugh, M.P., Johnson, C.P., and Tyler, T.F.Reliability of shoulder range of motion comparing a goniometer to a digital level. *Physiother Theory Pract* 26: 327–333, 2010.
144. Muyor, J.M., Vaquero-Cristobal, R., Alacid, F., and Lopez-Minarro, P.A.Criterion-related validity of sit-and-reach and toe-touch tests as a measure of hamstring extensibility in athletes. *J Strength Cond Res* 28: 546–555, 2014.
145. Naish, J.M. and Apley, J."Growing pains": a clinical study of non-arthritic limb pains in children. *Arch Dis Child* 26: 134–140, 1951.
146. Nakamura, M., Ikezoe, T., Takeno, Y., and Ichihashi, N.Effects of a 4-week static stretch training program on passive stiffness of human gastrocnemius muscle-tendon unit in vivo. *Eur J Appl Physiol*, 2011.
147. Nascimento, R., Desiree, M., Monteiro, E.R., Ribeiro, A., Reis, N., Sant'Ana, L., Vianna, J., Novaes, J., and Brown, A.Acute effect of different stretching methods in classical dancer children. *Brazilian Journal of Exercise Physiology* 19: 114–123, 2020.
148. Nelson, A.G. and Kokkonen, J.Acute ballistic muscle stretching inhibits maximal strength performance. *Res Q Exerc Sport* 72: 415–419, 2001.
149. Nuhu, A., Jelsma, J., Dunleavy, K., and Burgess, T.Effect of the FIFA 11+ soccer specific warm up programme on the incidence of injuries: A cluster-randomised controlled trial. *PLoS One* 16: e0251839, 2021.
150. O'Driscoll, S.L. and Tomenson, J.The cervical spine. *Clin Rheum Dis* 8: 617–630, 1982.
151. O'Hora, J., Cartwright, A., Wade, C.D., Hough, A.D., and Shum, G.L.Efficacy of static stretching and proprioceptive neuromuscular facilitation stretch on hamstrings length after a single session. *Journal of strength and conditioning research/ National Strength & Conditioning Association* 25: 1586–1591, 2011.
152. Osolin, N.G. *Das Training des Leichtathleten*. Berlin: Sportverlag, 1952.
153. Osolin, N.G. *Sovremennaia systema sportnnoi trenirovky* [*Athlete's training system for competitions*]. Moscow: Phyzkultura i sport, 1971.
154. Osternig, L., Robertson, R., Troxel, R., and Hansen, P.Differential responses to proprioceptive neuromuscular facilitation (PNF) stretch techniques. *Medicine and Science in Sports and Exercise* 22: 106–111, 1990.
155. Owoeye, O.B., Akinbo, S.R., Tella, B.A., and Olawale, O.A.Efficacy of the FIFA 11+ Warm-Up Programme in Male Youth Football: A Cluster Randomised Controlled Trial. *J Sports Sci Med* 13: 321–328, 2014.
156. Paradisis, G.P., Pappas, P.T., Theodorou, A.S., Zacharogiannis, E.G., Skordilis, E. K., and Smirniotou, A.S.Effects of Static and Dynamic Stretching on Sprint and Jump Performance in Boys and Girls. *Journal of Strength and Conditioning Research* 28: 154–160, 2014.
157. Patterson, P., Wiksten, D.L., Ray, L., Flanders, C., and Sanphy, D.The validity and reliability of the back saver sit-and-reach test in middle school girls and boys. *Res Q Exerc Sport* 67: 448–451, 1996.
158. Perrier, E.T., Pavol, M.J., and Hoffman, M.A.The acute effects of a warm-up including static or dynamic stretching on countermovement jump height, reaction time, and flexibility. *Journal of Strength and Conditioning Research/National Strength & Conditioning Association* 25: 1925–1931, 2011.
159. Perusse, L., Leblanc, C., and Bouchard, C.Familial resemblance in lifestyle components: results from the Canada Fitness Survey. *Can J Public Health* 79: 201–205, 1988.
160. Perusse, L., Leblanc, C., and Bouchard, C.Inter-generation transmission of physical fitness in the Canadian population. *Can J Sport Sci* 13: 8–14, 1988.

161. Polsgrove, M.J., Eggleston, B.M., and Lockyer, R.J.Impact of 10-weeks of yoga practice on flexibility and balance of college athletes. *Int J Yoga* 9: 27–34, 2016.
162. Power, K., Behm, D., Cahill, F., Carroll, M., and Young, W.An acute bout of static stretching: effects on force and jumping performance. *Med Sci Sports Exerc* 36: 1389–1396, 2004.
163. Pratt, M.Strength, flexibility, and maturity in adolescent athletes. *Am J Dis Child* 143: 560–563, 1989.
164. Radford, J.A., Burns, J., Buchbinder, R., Landorf, K.B., and Cook, C.Does stretching increase ankle dorsiflexion range of motion? A systematic review. *Br J Sports Med* 40: 870–875; discussion 875, 2006.
165. Roaas, A. and Andersson, G.B.Normal range of motion of the hip, knee and ankle joints in male subjects, 30–40 years of age. *Acta Orthop Scand* 53: 205–208, 1982.
166. Russell, B., Dix, D.J., Haller, D.L., and Jacobs-El, J.Repair of injured skeletal muscle: a molecular approach. *Med Sci Sports Exerc* 24: 189–196, 1992.
167. Saal, J.S.Flexibility Training. In: *Functional Rehabilitation of Sports and Musculoskeletal Injuries.* Gaithersburg, MD: Aspen, 1998, pp. 84–97.
168. Sady, S.P., Wortman, M., and Blanke, D.Flexibility training: ballistic, Static or proprioceptive neuromuscular facilitation? *Archives of Physical Medicine and Rehabilitation* 63: 261–263, 1982.
169. Safran, M., Garrett, W., Seaber, A., Glisson, R., and Ribbeck, B.The role of warmup in muscular injury prevention. *The American Journal of Sports Medicine* 16: 123–129, 1988.
170. Safran, M., Seaber, A., and Garrett, W.Warm-up and muscular injury prevention. An update. *Sports Medicine* 8: 239–249, 1989.
171. Sahrmann, S.A. *Diagnosis and treatment of movement impairment syndromes.* St. Louis, MS: Mosby Publishers Inc., 2002.
172. Sale, D. and MacDougall, J.D.Specificity in Strength Training: A Review For The Coach and Athlete. *Can J App Sports Sci* 6: 87–92, 1981.
173. Salminen, J.J.The adolescent back. A field survey of 370 Finnish schoolchildren. *Acta Paediatr Scand Suppl* 315: 1–122, 1984.
174. Samuel, M.N., Holcomb, W.R., Guadagnoli, M.A., Rubley, M.D., and Wallmann, H.Acute effects of static and ballistic stretching on measures of strength and power. *J Strength Cond Res* 22: 1422–1428, 2008.
175. Seguin, R.A., Heidkamp-Young, E., Kuder, J., and Nelson, M.E.Improved physical fitness among older female participants in a nationally disseminated, community-based exercise program. *Health Educ Behav* 39: 183–190, 2012.
176. Sekir, U., Arabaci, R., Akova, B., and Kadagan, S.M.Acute effects of static and dynamic stretching on leg flexor and extensor isokinetic strength in elite women athletes. *Scand J Med Sci Sports* 20: 268–281, 2010.
177. Sharman, M.J., Cresswell, A.G., and Riek, S.Proprioceptive neuromuscular facilitation stretching: mechanisms and clinical implications. *Sports Med* 36: 929–939, 2006.
178. Sherrington, C.S.Flexion-reflex of the limb, crossed extension reflex stepping and standing. *J Physiol* 40: 28–121, 1910.
179. Sherrington, C.S.Remarks on some aspects of reflex inhibition. *Proc Royal Soc Lond* 97: 519–519, 1925.
180. Silvers-Granelli, H., Mandelbaum, B., Adeniji, O., Insler, S., Bizzini, M., Pohlig, R., Junge, A., Snyder-Mackler, L., and Dvorak, J.Efficacy of the FIFA 11+ Injury

Prevention Program in the Collegiate Male Soccer Player. *Am J Sports Med* 43: 2628–2637, 2015.

181. Slauterbeck, J.R., Choquette, R., Tourville, T.W., Krug, M., Mandelbaum, B.R., Vacek, P., and Beynnon, B.D.Implementation of the FIFA 11+ Injury Prevention Program by High School Athletic Teams Did Not Reduce Lower Extremity Injuries: A Cluster Randomized Controlled Trial. *Am J Sports Med* 47: 2844–2852, 2019.
182. Smith, C.The warm-up procedure: to stretch or not to stretch. a brief review. *Journal of Orthopedic Sports Physical Therapy* 19: 12–17, 1994.
183. Soucie, J.M., Wang, C., Forsyth, A., Funk, S., Denny, M., Roach, K.E., Boone, D., and Hemophilia Treatment Center Network. Range of motion measurements: reference values and a database for comparison studies. *Haemophilia* 17: 500–507, 2011.
184. Spernoga, S.G., Uhl, T.L., Arnold, B.L., and Gansneder, B.M.Duration of Maintained Hamstring Flexibility After a One-Time, Modified Hold-Relax Stretching Protocol. *J Athl Train* 36: 44–48, 2001.
185. Stone, W.J. and Kroll, W.A. *Sports conditioning and weight training: programs a for athletic competion*. Dubuque, IA: Wm. C. Brown, 1991.
186. American Academy of Orthopaedic Surgeons. *Joint motion: methods of measuring and recording*. Edinburgh: E&S Livingstone, 1969.
187. Thomas, E., Bianco, A., Paoli, A., and Palma, A.The Relation Between Stretching Typology and Stretching Duration: The Effects on Range of Motion. *Int J Sports Med* 39: 243–254, 2018.
188. Tippett, S.R.Lower Extremity Strength and Active Range of Motion in College Baseball Pitchers: A Comparison between Stance Leg and Kick Leg. *J Orthop Sports Phys Ther* 8: 10–14, 1986.
189. Tran, M.D., Holly, R.G., Lashbrook, J., and Amsterdam, E.A.Effects of Hatha Yoga Practice on the Health-Related Aspects of Physical Fitness. *Prev Cardiol* 4: 165–170, 2001.
190. van de Pol, R.J., van Trijffel, E., and Lucas, C.Inter-rater reliability for measurement of passive physiological range of motion of upper extremity joints is better if instruments are used: a systematic review. *J Physiother* 56: 7–17, 2010.
191. van Trijffel, E., van de Pol, R.J., Oostendorp, R.A., and Lucas, C.Inter-rater reliability for measurement of passive physiological movements in lower extremity joints is generally low: a systematic review. *J Physiother* 56: 223–235, 2010.
192. Wallin, D., Ekblom, B., Grahn, R., and Nordenberg, T.Improvement of muscle flexibility: a comparison between two techniques. *The American Journal of Sports Medicine* 13: 263–267, 1985.
193. Whatman, C., Knappstein, A., and Hume, P.Acute changes in passive stiffness and range of motion post-stretching. *Physical Therapy in Sport* 7: 195–200, 2006.
194. Wing, P., Tsang, I., Gagnon, F., Susak, L., and Gagnon, R.Diurnal changes in the profile shape and range of motion of the back. *Spine (Phila Pa 1976)* 17: 761–766, 1992.
195. Wyon, M.A., Smith, A., and Koutedakis, Y.A comparison of strength and stretch interventions on active and passive ranges of movement in dancers: a randomized controlled trial. *J Strength Cond Res* 27: 3053–3059, 2013.
196. Youdas, J.W., Krause, D.A., Hollman, J.H., Harmsen, W.S., and Laskowski, E. The influence of gender and age on hamstring muscle length in healthy adults. *J Orthop Sports Phys Ther* 35: 246–252, 2005.
197. Young, W. and Behm, D.Should static stretching be used during a warm-up for strength and power activities? *Strength and Conditioning Journal* 24: 33–37, 2002.

198. Young, W.B.The use of static stretching in warm-up for training and competition. *IntJSports Physiol Perform* 2: 212–216, 2007.

199. Young, W.B. and Behm, D.G.Effects of running, static stretching and practice jumps on explosive force production and jumping performance. *J Sports Med Phys Fitness* 43: 21–27, 2003.

200. Zito, M., Driver, D., Parker, C., and Bohannon, R.Lasting effects of one bout of two 15-second passive stretches on ankle dorsiflexion range of motion. *J Orthop Sports Phys Ther* 26: 214–221, 1997.

4 Mechanisms Underlying Acute Changes in Range of Motion

Thixotropic Effects

In fact, it is not even necessary to stretch in order to temporarily increase range of motion (ROM). Warming the muscles and tendons will easily improve your elastic flexibility. Even more effective is to include an aerobic activity or warm-up muscle contractions. The traditional warm-up is typically initiated with a submaximal aerobic component (e.g., running, cycling) to raise the body temperature 1–2°C (121, 122). Any study that measures joint ROM before and after almost any activity that involves some persistent muscle contractions will detect an increase in ROM. S. Peter Magnusson: a well-respected stretch researcher from Denmark indicated that the acute effects of stretching in the holding phase of a stretch are due to changes in tissue visco-elasticity. The underlying mechanism for this visco-elastic effect is thixotropy. Thixotropy occurs when viscous (thicker) fluids become less viscous or more fluid-like when agitated, sheared, or stressed. When the stress is removed, or desists then the fluid takes a certain period to return to the original viscous state. Muscle contractions are not very efficient. Only 40–60 percent of the energy consumed during a contraction contributes to producing force (i.e., myofilament, Ca^{++}, and Na^+/K^+ pump kinetics), whereas 40–60 percent is released as heat (105). The muscle contraction-induced increase in temperature of the soft tissues can decrease the viscosity of intracellular and extracellular fluid providing less resistance to movement. Increases in muscle temperature can occur with higher environmental temperatures, and muscle contractions associated with dynamic stretching movements, isometric contractions during PNF stretching and to a lesser extent with the reflexive contractions of static stretching. Hamstrings passive static stretching elevated the local skin temperature by approximately 0.3°C after 180 seconds of stretching (32). Thixotropic effects on viscosity are not just muscle-related, but as tendons consist of 55–70 percent water (62), they will also be affected by viscosity changes.

A great analogy for those individuals who live in northern climates is the viscosity of the oil in the car or truck engine when it is extremely cold. If you imagine that the pistons are myosin molecules, both have to move in order to create movement. Oil in very cold temperatures would act more like viscous

DOI: 10.4324/9781032709086-5

molasses providing high resistance to the movement of the pistons. A cold muscle would have more viscous sarcoplasm providing higher resistance to the intramuscular proteins such as myosin, titin and others. The extracellular fluid would also be more viscous when cold and thus provide more resistance to the movement or sliding of muscle fibres, tendons and fascia. If a car is cold and will not start, you plug in the block heater which can warm up the oil (decrease viscosity) so the pistons can move. If your muscles are cold, muscle contractions can increase musculotendinous temperatures decreasing viscosity and resistance to movement. In fact, I have had lazy friends who rather than stretch and actively warm up before a squash game, would sit in the sauna to warm their bodies and thus "limber up" (i.e., decrease their musculotendinous and myofascial resistance to movement). Athletes playing in cold environments must keep this concept in mind. The colder they become, less pliability and less flexibility will result. Often, you may see or hear an athlete who will disdain from standing by a heater or using a blanket or jacket while on the sidelines in a cold or freezing environment. Their intended approach or message is that they are so psychologically "tough" they do not need to use the same devices as some "soft" or psychologically "weaker" opponents. As a scientist and a coach, I would rather that my athletes stay warm and be psychologically "smart" and physiologically "efficient", as hypothermia (cold) not only affects resistance to movement (flexibility), but strength, power, rate of force development, endurance, metabolism, and other vital processes for success in the event or sport (36, 41).

There are a number of other explanations for the increased ROM immediately after stretching. Depending on whether the stretching is static, PNF, or dynamic, there may be various emphases on whether it is thixotropic, neural, mechanical, or psychological (stretch tolerance).

Neural Mechanisms of Acute Static Stretching

Although generally we try to relax when performing passive static stretches, there is still some muscle activation. Blazevich et al. (19) recorded electromyography (EMG) activity (normalized to maximal voluntary isometric contraction − MVC) of 8.5 percent (soleus) and 5.5 percent (medial gastrocnemius) at the point of maximal stretch, thus suggesting that muscle reflexes would likely have contributed to the passive stretch resistance of the musculotendinous unit. They split their participants into flexible and inflexible individuals and found a moderate (0.60) correlation between the angle of EMG onset and maximum range of motion. Thus, more flexible people did not experience significant muscle activity (reflex-induced EMG) until they achieved a greater ROM. According to Nathalie Guissard and Jacques Duchateau (51), two internationally distinguished neuromuscular physiologists from Belgium, the amount of stretch or joint ROM that can be produced is highly attributable to the extent of muscle resistance caused by tonic reflexes. Whereas, dynamic movements (not through a full ROM) and dynamic stretching (a full or nearly full ROM) tend to excite the neuromuscular system, static stretching is purported to decrease or disfacilitate this reflexive activity excitation of the motoneurones (11, 12, 13, 15). The origin of this reflex suppression can be tested by using the Hoffmann reflex (H-reflex) and the tendon reflex (T-reflex) (Figure 4.1). The H-reflex is evoked by stimulation of the Ia fibres to monitor the afferent excitability of the alpha motoneurons (38). Decreases in the H-reflex amplitude can signal decreases in motor neuron excitability or presynaptic inhibition (inhibition of interneuron (s) innervating Ia terminals) of the Ia afferents. What is the difference between inhibition and disfacilitation? Inhibition is a more active process caused by the activation of inhibitory interneurons leading to inhibitory postsynaptic potentials in the motoneurons (112). These interneurons would release inhibitory neurotransmitters such as y-aminobutyric acid (GABA). However, with disfacilitation, motoneurons would be hyperpolarized due to the temporal absence of excitatory synaptic activity (112). In simpler terms, if you were riding a bicycle and you slowed your pedaling rate or stopped rotating the crank, you would decrease the activity and the speed of the bicycle would be disfacilitated. However, if you applied the brake pads to the wheels then you would actively attempt to slow the bike just like an interneuron and thus you would be directly inhibiting the bike's speed. The T-reflex could be a combination of inhibition or disfacilitation. T-reflex is elicited by tapping

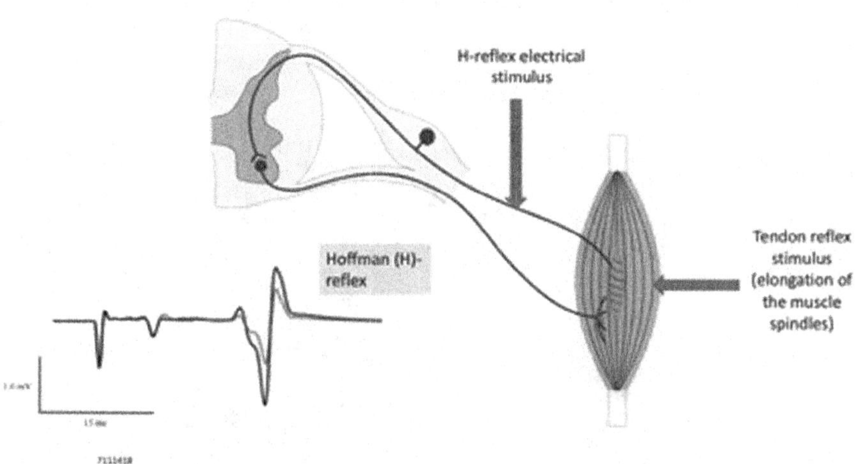

Figure 4.1 Hoffman (H-) reflex and tendon reflex. The stimulus for the H-reflex is a low-intensity electrical stimulation sufficient to activate the afferents to the spinal motoneurons but minimally activating the alpha motoneuron efferents so as not to elicit a strong muscle action potential (M-wave). The stimulus for tendon reflex is an elongation of the muscle to excite or activate the nuclear bag and chain fibres of the intrafusal muscle spindles to elicit a myotactic reflex

(percussion) of the tendon and could be affected by central changes (disfacilitation of motor neuron excitability and presynaptic inhibition) and by changes in the muscle spindle sensitivity/activity (disfacilitation). One hour of repeated passive stretching of the plantar flexors has been shown to decrease the stretch reflex by 85 percent while the H-reflex was reduced by 44 percent (7). However, you would have to ask, who would ever stretch their calf muscles for an hour. A good example how lab research which tries to understand mechanisms does not always parallel real life.

With passive static stretching, the muscle is typically extended at a slow to moderate rate into an elongated position by another person or a device (i.e., rubber band or machine) (see Figure 4.1). This extended or elongated position is then held for an extended period of time, typically from 15–60 seconds (3, 12). It is the responsibility of the muscle spindles to detect, continuously monitor and send signals to the central nervous system regarding the rate and extent of elongation of the muscle (80, 81). It can be regarded as both a protective system and a proprioceptive (position sense) system. As a protective system, if the muscle is elongated at a rapid rate to the end or near the end of the ROM then a reflexive signal will be sent to the spinal cord to contract the muscle that is being stretched (myotactic stretch reflex: see Figure 4.2) (82). This reflex is the action of the quadriceps muscle that you experience when a doctor taps your patellar tendon (front of the knee) and then your lower leg jerks forward (T-reflex). The doctor's hammer has elongated the patellar tendon which is attached to the quadriceps; stretching that muscle, which was detected by the muscle spindles leading to a reflex to the spinal cord, causes your

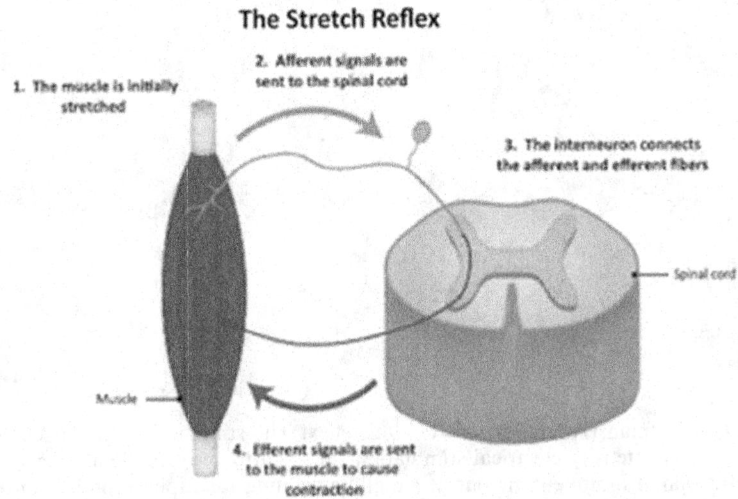

Figure 4.2 Tendon or myotactic reflex with monosynaptic excitation of the agonist muscle and di-synaptic reciprocal inhibition of the antagonist muscle (reciprocal inhibition not shown in this figure)

quadriceps to contract. This reflex contraction of the elongated muscle should therefore help prevent the joint skeletal structures (bones and cartilage) and ligaments from exceeding the joint and muscle's functional ROM and thus helping to prevent damage or injury. While the elongated muscle is being reflexively contracted the antagonist to that muscle is being inhibited. The myotactic reflex is monosynaptic (one synapse), whereas the antagonist inhibition is di-synaptic (two synapses). In terms of movement, this antagonist inhibition is efficient as it automatically reduces muscle contractions that work in opposition to the intended motion. Thus, if you are walking, the dynamic contraction and elongation of the quadriceps will lead to further excitation of the quadriceps, which are propelling you along while the hamstrings are being inhibited to reduce the resistance to the walking movement. However, if you are sprinting and the hamstrings get stretched or elongated to a great extent, they also will be subjected to myotactic reflexes to protect them from being extended too far.

Within the muscle spindle, there are nuclear chain fibres that preferentially respond to changes in the amount of stretch or elongation of the muscle whereas the nuclear bag fibres respond to both the extent and rate of elongation (see Figure 4.3). When activated, nuclear chain and bag fibre impulses are sent to the spinal motoneurons via annulospiral (high conduction velocity) and flower spray endings (low conduction velocity) through Ia (nuclear bag and chains) and II (nuclear chains) afferents respectively. As previously mentioned, a monosynaptic (just one synapse) myotactic reflex is initiated from the rapid stretch action resulting in the depolarization/activation (contraction) of the alpha motoneuron of the stretched muscle and inhibition (di-synaptic) of the antagonist muscle (82).

The spindles also work to inform the system about a body segment's position in space or proprioception. The intensity or discharge rate of the impulses from the muscle spindles will inform the central nervous system about the

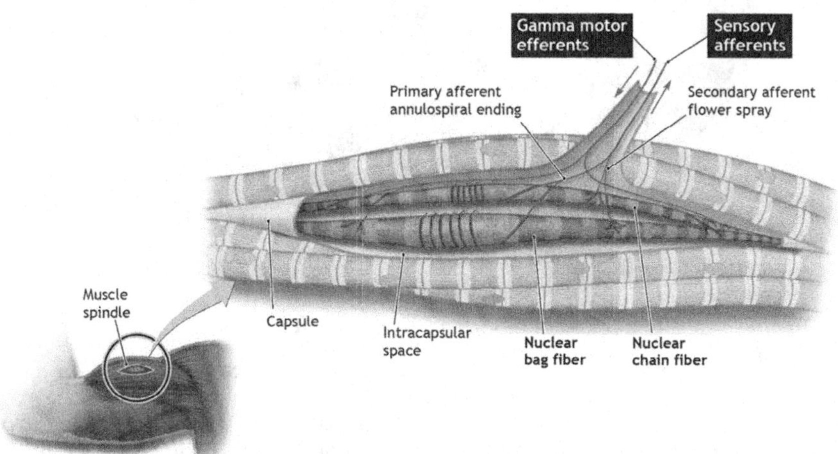

Figure 4.3 Nuclear bag and chain fibres are located within an intrafusal muscle spindle

rate and extent of movement (97). That is why you can close your eyes, abduct and extend your arms and then still touch your nose. Your muscle spindles are informing the system about how fast and far you have moved in order to reach your nose and based on prior experience, when to slow down and stop so you do not break your nose.

Another receptor that is activated by tension exerted on the muscle tendon and muscle is the Golgi tendon organ (GTO) (26, 57). The GTO receptors within the tendon are shaped like coiled elastics and when they are uncoiled due to tension, a piezo-electric effect occurs (see Figure 4.4). Certain biological materials such as crystals (bone), proteins, DNA and others can discharge an electric charge in response to mechanical stress (102). Piezoelectricity means electricity resulting from pressure, derived from the Greek piezō or piezein, which means to squeeze or press. So, when a contraction occurs or

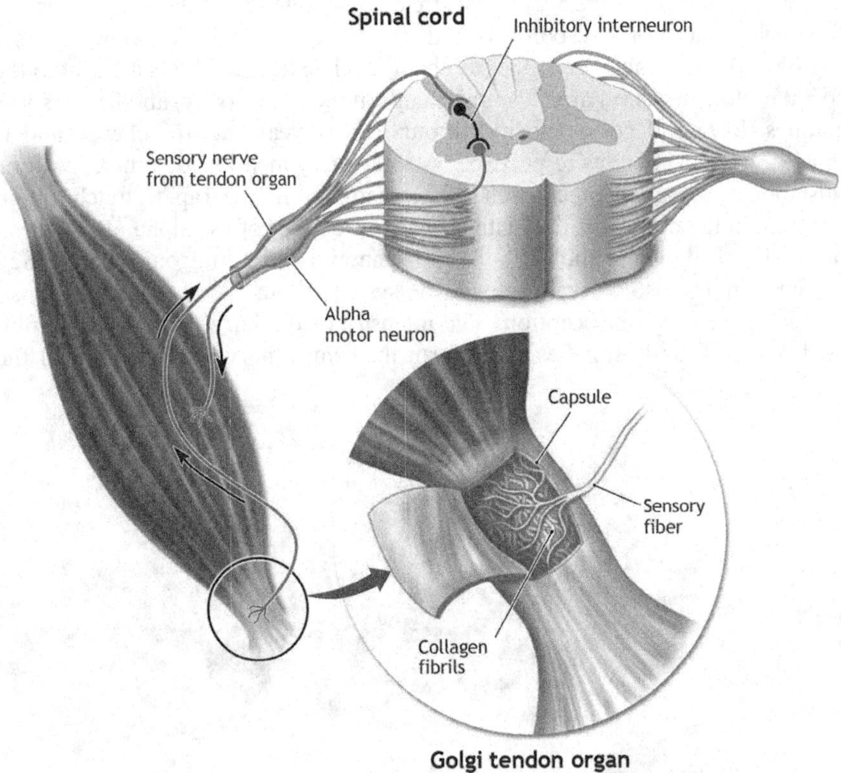

Figure 4.4 Golgi tendon organs (GTO) are located in the muscle-tendon or aponeurosis-tendon junction and they can lie in series or in parallel with the extrafusal muscle fibres, so they can detect force increases not only from fibres in series with the receptor but distant to the GTO through the parallel connections

tension is exerted on the musculotendinous unit, the GTO discharges signals to the central nervous system. While the GTO can reflexively excite the system, under these circumstances, it typically results in an inhibitory signal that is labelled as an autogenic inhibition. This reflex is di-synaptic in that the GTO synapses with an inhibitory interneuron in the spinal cord, which then inhibits the same muscle group that experienced the tension (61). Although theoretically, the GTO should contribute to stretch-induced reflex inhibition, Edin and Vallbo (37) found that while most spindle afferents respond rapidly to stretch, GTO were insensitive to the tension produced on the tendon with stretch. If there is a stretch-induced GTO inhibition, it is more likely to occur with large amplitude stretches (51). Furthermore, any possible GTO inhibition subsides almost immediately (60–100 milliseconds post-stretching) after the stretching discontinues (57).

In conjunction with the GTO inhibition to large amplitude stretches, the Renshaw cells can also play a minor role with large amplitude stretches. Renshaw cell inhibition is also known as recurrent inhibition (see Figure 4.5). Recurrent or Renshaw cell inhibition can exert stabilizing effects on motoneurone discharge variability, and motor unit synchronization during voluntary muscle contractions (79). Recurrent inhibition is more prevalent with weak rather than strong contractions, and phasic rather than tonic contractions (59). As the stress or tension

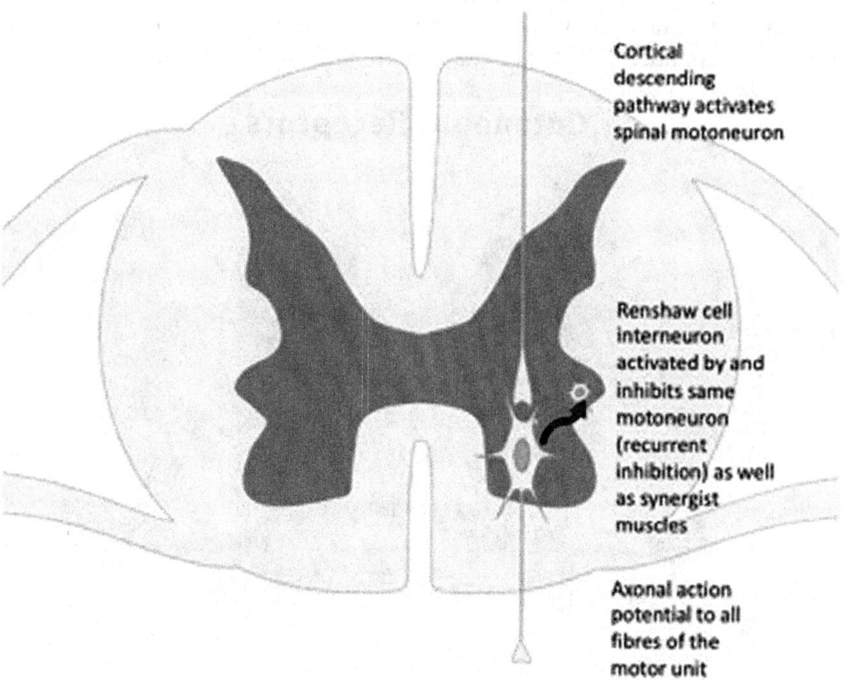

Figure 4.5 Renshaw cells and Recurrent Inhibition

on the muscle during a large-amplitude stretch is still much less than with a maximal contraction, it should contribute to motoneuron inhibition and muscle relaxation. However, as it is more prevalent with phasic contractions, it may play a more predominant role with full ROM dynamic stretching compared with lesser effects with static stretching.

We have discussed the possible roles of nuclear chain and bag fibres of the muscle spindles, Renshaw cells, and GTOs. Cutaneous nerve fibres can also contribute to the ROM capabilities (see Figure 4.6). Cutaneous receptors have polysynaptic innervations to motoneurons and are monitored with the exteroceptive (E-) reflex. Research has demonstrated that small amplitude passive stretches induce pre-synaptic inhibition (e.g., decreased H-reflex) but no change in the E-reflex (cutaneous) (34) or corticospinal excitability as measured with motor evoked potentials induced by transcranial magnetic stimulation. However, with a greater ROM, both H- and E-reflexes decreased similarly, suggesting that post-synaptic inhibitory mechanisms contribute to the observed changes, but they only persist for a few seconds following the completion of the stretching. According to Guissard and Duchateau, (51), joint and cutaneous receptors are not significant inhibitors with small amplitude stretches with only a small contribution during large amplitude stretches. Furthermore, increases in the afferent excitability of the motoneurons (H-reflex) are associated with increases in the musculotendinous passive torque (31).

With an active static stretch, the individual will use another muscle group – typically, the antagonist muscles to help move the target muscle and joint through an extended ROM. For example, to stretch the hamstrings, you may

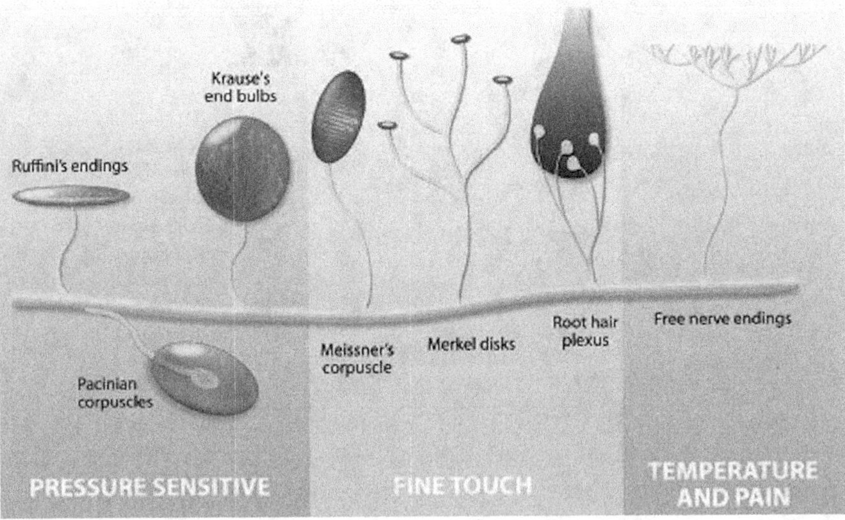

Figure 4.6 Cutaneous nerve receptors

actively contract the quadriceps to flex the hip which will naturally extend the hamstring muscles which work as hip extensors. Reciprocal inhibition (29, 95) may play a role in this situation, as the quadriceps contraction changes the length of the quadriceps muscles spindles at a particular rate and can lead to an inhibitory di-synaptic reflex synapse with its antagonist, leading to more relaxation in the hamstrings.

So, let's go through a static stretching scenario of a person who is lying on their back (supine position) and in the first instance, a partner flexes the hip of their extended leg towards their chest to stretch the hamstrings and in the second situation the person contracts their quadriceps and uses their arms to pull their extended leg toward their chest. In the passive static stretching situation of the first scenario, the muscles spindles, specifically, the nuclear bag fibres would detect a slow to moderate rate of muscle length change, while the nuclear chain fibres would predominately monitor the extent of change in muscle length. Both nuclear bag and chain fibres would discharge at a relatively high frequency with annulospiral and flower spray ending through the Ia and II afferent nerve trunks to initiate the myotactic reflexes and cause contractions of the hamstrings, thus resisting the stretching of the muscle. In one study, although subjects were instructed to relax during a dorsiflexors stretch, EMG activity reached 5.5 percent and 8.5 percent of MVC in the gastrocnemius and soleus muscles respectively (19). Therefore, muscle spindle reflexes would have contributed to the "passive" resistance of the musculotendinous unit during the stretch.

Naturally, once the stretch had reached the maximum ROM, there would no longer be any change in the rate of stretch. Within seconds, the nuclear bag fibres would begin to decrease their discharge rate, hence decreasing the intensity of the myotactic reflex contractions (disfacilitation). As the stretch was held in this position for a prolonged period (typically 15–60 seconds), the nuclear chain fibres would accommodate this new position and also decrease their discharge frequency (disfacilitation). This decreased discharge frequency can partially be attributed to the actions of the gamma efferent system (see Figure 4.3). The gamma efferent system attempts to return the muscle spindles to their reference length after or during movement (81). Just like the extrafusal muscle (skeletal muscle innervated by the alpha motoneuron), the intrafusal muscles (location of sensory muscle spindles) also have contractile myofilament proteins. Thus, the gamma efferent system can activate these myofilaments leading to a spindle contraction that would shorten the length of the spindle. By decreasing the length of the spindle, the nuclear chain fibres decrease their discharge rate and the myotactic contractions decrease further, leading to more relaxation of the muscle. Although elongating the muscle to the end of ROM would have placed tension on the GTO that should activate greater type Ib afferent inhibitory stimulation there is little evidence that it contributes to substantial muscle relaxation (perhaps a small contribution with an extensive ROM to maximal point of discomfort). Try this yourself. Stretch any muscle passively and hold it for 30 seconds or more. Then see if you can stretch the muscle or increase the ROM further. It is guaranteed that you will have a greater ROM. If you used an active static stretch then the contraction of the quadriceps may have also contributed to further relaxation with the reciprocal inhibition effect.

So, which process is more predominant with static stretching: inhibition or disfacilitation? In one study H- and T-reflexes were compared in the soleus with increases in ankle ROM (52). The H-reflex amplitude decreased more than the T-reflex (31 percent vs. 8 percent of control). When the stretching was completed the H-reflex returned immediately to its control value, while the T reflex remained depressed. Therefore, the T-reflex reduction was more likely derived from either a decrease in muscle spindle sensitivity (disfacilitation) or increased musculotendinous unit compliance (mechanical change). According to this study, during the stretch there could be pre-synaptic inhibition and muscle spindle-induced disfacilitation of the motoneuron, whereas the persistent increases in ROM would be attributed more to spindle disfacilitation and perhaps musculotendinous compliance. H-reflex inhibition has been reported to have two recovery stages during 30 seconds of stretching. It was hypothesized that H-reflex inhibition present until 18 seconds was related to post-activation depression of Ia afferents caused by the passive dorsiflexion movement. From 21 seconds, the lower inhibition may have been caused by a weaker post-activation depression, and inhibition from secondary afferents or post-synaptic inhibition (23).

Whilst the H-reflex can be inhibited during the stretch, this H-reflex inhibition is short lived following the stretch (22). This rapid recovery has been suggested to be influenced by a transient increase in corticospinal excitability (22). Further evidence for increased cortical excitability was provided by Opplert et al. (94) who found increased corticospinal excitability but a lack of change in spinal excitability after both static and dynamic muscle activity.

Budini et al. (24) reported that tendon tap reflexes could be inhibited for ten minutes after 30 seconds of static stretching. They demonstrated that it was possible to promote an immediate recovery from this spindle inhibition or disfacilitation and restore reflex sensitivity by performing a maximal voluntary isometric contraction. However, three dynamic contractions delayed the spindle reflex recovery by 90 seconds.

In fact, we can go even deeper than that. Are there differences in how the muscle reflexes are inhibited if you have a small versus large amplitude stretch? It has been shown that passive small amplitude stretches decrease the H-reflex through pre-synaptic inhibition. Pre-synaptic Ia inhibition pathway originates from the intrafusal muscle spindle fibres and projects pre-synaptically (before the junction of the dendrite and the motoneuron) by interneurons onto the Ia terminals (53). Thus, there is no intrinsic change in motoneuron excitability. With larger amplitude passive stretches, there may be a suppression of the cortical neurones and ⊠ motoneuron excitability (53). In this case with large amplitude stretches, GTO inhibitory afferents could contribute to decreased motoneuron excitability from Ib fibres (53). GTO respond primarily to muscle contraction forces, but are not very sensitive to the degree of force or stress associated with smaller amplitude passive stretching (53). Remember though, that GTO effects would only continue during the period of stretch and tend to diminish rapidly thereafter (57). There is also the possibility of post-synaptic inhibition with large amplitude stretches from the inhibitory effect of the Renshaw cell recurrent inhibition loop (25).

In fact, it is possible, that stretch-induced reflex interneuron inhibition might affect not only the muscles being stretched but distant or non-local muscles as well. There are a few studies that have demonstrated that stretching the hamstrings unilaterally (only one side of the body) will improve ROM of the contralateral (opposite side) hamstrings (27). Our lab has also shown that static or dynamic stretching of the shoulders will increase ROM of the hamstrings, while stretching the hip adductor (groin) muscles increased the flexibility of the shoulders (14). This research could be crucial evidence for the pervasive effects of stretch-induced neural inhibition that acts globally on the body.

Neural Mechanisms of Acute Proprioceptive Neuromuscular Facilitation Stretching

Traditionally, Proprioceptive Neuromuscular Facilitation (PNF) was considered more effective than static stretching for increasing ROM due to the variety of reflex inhibition techniques that were proposed to be involved. For example, with the CRAC (contract-relax-agonist-contract) method, the muscle to be stretched (e.g., hamstrings) would be actively placed in an elongated position by contracting the antagonist muscle (e.g., hip flexion by the quadriceps). The contraction of the antagonist would activate reciprocal inhibition. Reciprocal inhibition is a reflex branch of the myotactic reflex. As a reminder, the myotactic reflex is a monosynaptic reflex initiated from the stretching of the muscle spindles resulting in an excitation of the stretched muscle. This muscle spindle-induced excitation also has an afferent nerve branch that excites an inhibitory interneuron that suppresses the activity of an antagonist motoneuron (di-synaptic). In this example, contracting the quadriceps to lengthen the hamstrings would change the length of many quadriceps muscle spindles (especially in the distal regions), causing further excitation of the quadriceps with inhibition of the hamstrings permitting greater muscle excursion (elongation). The next step with CRAC PNF stretching would be to contract the stretched muscle (e.g., hamstrings) against a partner's resistance or it can also be against an immovable object like a wall. This contraction, which does not need to be a maximal contraction, as Roger Enoka (a very prominent and internationally renowned neuromuscular scientist) and colleagues (38) found that whether the contraction was performed at 50 or 100 percent of the subjects' maximal voluntary force, there was a similar reduction in the H-reflex amplitude (afferent excitability of the alpha motoneuron). Furthermore, this contraction of an elongated muscle would place stress on the GTO, possibly leading to its activation inducing autogenic inhibition (56). Autogenic inhibition from GTO activation increases Ib muscle afferent activity (57). Ib afferent activity hyperpolarize the dendrites synapsing with spinal alpha motoneurons of the stretched muscle, decreasing or blocking the Ia afferent reflex activity enabling further increases in ROM (26). However, there is no direct evidence for a positive relationship between the aforementioned reflex activity and PNF ROM (26). One might expect that this reciprocal and autogenic reflex inhibition would decrease muscle activity as evidenced by attenuated EMG activity, but a number of studies have illustrated increased resting EMG activity immediately after the contraction phase of a PNF stretch (71, 87). As GTO effects persist briefly after the tension to the receptors is removed (113), the autogenic inhibition effects should not persist and its contribution to PNF flexibility is highly debatable (56, 103). Although, a depression of the spinal reflex excitability after the isometric contraction is brief (<5 seconds), it could still persist long enough to provide an advantage for the subsequent stretch. Thus, similar to static stretching, there could be some reflex inhibition during the stretching procedure. An increased stretch tolerance (86), decreased visco-elasticity and a degree of reduced musculotendinous stiffness (60, 71) could all contribute to the sustained increase in elastic ROM.

Neural Mechanisms of Acute Dynamic Stretching

Whereas static and PNF stretching should reduce muscle activation through some degree of reflexive disfacilitation with reduced nuclear bag and nuclear chain muscle spindle receptor activity (Ia afferents) with further possible inhibition from autogenic (Ib afferents) and reciprocal inhibition (Ia afferents), dynamic stretching should excite or increase activation of the system. The previously discussed myotactic reflex activity would be increased with dynamic stretching due to nuclear bag and chain excitation from the higher rate and extent of muscle elongation respectively. As dynamic muscle stretching is usually performed as a relaxed action with submaximal muscle contractions, the activation of GTOs initiating autogenic inhibition would not be expected to play a major role in increasing acute ROM. Reciprocal inhibition would be activated with the sequential movement of the limbs similar to the well documented reciprocal inhibition sequences found with locomotor activities such as walking and running (96, 98). This reciprocal inhibition could contribute to greater dynamic movement excursions during the stretching activity but would not persist after the activity. Hence, for the augmentation of ROM to persist after the dynamic stretching, the reciprocal inhibition would need to contribute to viscous and morphological changes, that would continue for a prolonged period after the stretching. Typically, dynamic stretching is not as uncomfortable or painful as static or PNF stretching, hence the role of stretch tolerance may not be as predominant.

Stretch Tolerance Sensory Theory

Increases in ROM could also be attributed to a psycho-physiological effect (sensory theory). Magnusson and colleagues suggest that increased flexibility can be largely attributed to an increased stretch tolerance (72, 73). They showed that after three weeks of stretch training, the increased ROM was not attributed to differences in muscle stiffness or reflex EMG activity (72). Increased stretch tolerance could be related to changes in the sensitivity of nociceptive (pain) nerve endings (77) that allow the individual to accommodate greater discomfort or pain and thus push themselves through a greater ROM. Stove and colleagues (106) induced remote pain (cold pressor test of non-dominant hand) intended to elicit endogenous pain inhibitory systems and found an increase in knee extension ROM. They suggested there was a link between increased stretch tolerance and endogenous pain inhibition. Diffuse noxious inhibitory control theory indicates that activity of A delta- or A delta- and C-peripheral fibres influence convergent neurons in the spinal dorsal horn resulting in a global release of endorphins and enkephalins (morphine-like compounds) that promote analgesic effects allowing the individual to sustain greater discomfort or pain and stretch further (65). Enhanced stretch tolerance rather than changes in passive muscle stiffness has been demonstrated with seniors regardless of sex (91). Just 20 seconds of static stretching increased ROM but did not affect the shear elastic modulus (muscle elasticity) again pointing to an increase in stretch tolerance as the mechanism (100). Their ROM improvement persisted for 10 minutes. Passive stretching can produce acute ischemic compression, which has been shown to result in reduced perceived pain in the neck and shoulder muscles (54). Freitas et al. (45) in a meta-analysis examined 26 stretch training studies that ranged from 3–8 weeks in duration with a weekly stretching duration of 1,165 seconds per week (about 20 minutes). Stretching for 3–8 weeks did not on average alter muscle and tendon properties and thus the increased extensibility must have been related to a greater tolerance to tension. However, if stretch tolerance was the only mechanism at play with no substantial reflex inhibition occurring, then the commonly found static stretch-induced performance impairments would not be expected from non-stretched muscles.

But we found that not only can stretching one muscle increase the ROM of another muscle but we have also reported that unilateral static stretching of the plantar flexors (calf muscles) led to impairments of jump height in the contralateral lower limb (30) and stretching the shoulders also impaired lower limb jump performance (75). Static stretching of the pectoralis major muscle decreased activation of the triceps brachii during a bench press action (76). There are not many questions in the world that are simply black and white. So, while an increased stretch tolerance (totally psychological or psycho-physiological influenced by type III and IV nociceptor afferents) can certainly help explain ROM improvements, it seems likely that some degree of afferent stretch-induced reflex inhibition is also acting on the muscle under stretch as well as other non-affected muscles. While this reflex inhibition would be most predominant during the holding of the stretch, the reported subsequent impairments in non-local jump performance and muscle activation suggests that it can persist for a few minutes following stretching. However, even, if neural reflex inhibition works primarily during the actual stretching exercise permitting a greater elongation of the muscle, then it might impact morphological structures such as muscles, tendons and in some cases ligaments.

Acute Morphological Static Stretching Mechanisms

Skeletal Structures

Greater stretch tolerance or neural inhibition should allow the muscle to be elongated to a greater degree. Maintaining this greater elongation over an extended period (i.e., 20–60 seconds) might be expected to affect the properties of the musculotendinous tissues. What musculoskeletal components restrict our ROM. ROM is affected by skeletal structures, joint capsules, fascia, ligaments, muscles, tendons, aponeuroses, and fat. What factors can we modify with stretching? If we stretch till a bone is fractured, we can get an acute increase in ROM but at what cost: excessive pain, inflammation, and loss of function. Within moments, the pain and inflammation will then decrease the ROM! The glenoid fossa or cavity of the shoulder (glenohumeral) joint is a relatively flat surface allowing the shoulder a great deal of unrestricted movement for flexion/extension, abduction/adduction, horizontal abduction/adduction, medial and lateral rotation, and circumduction (see Figure 4.7). In contrast, the acetabulum of the hip joint is deeper and more cup-like restricting ROM compared

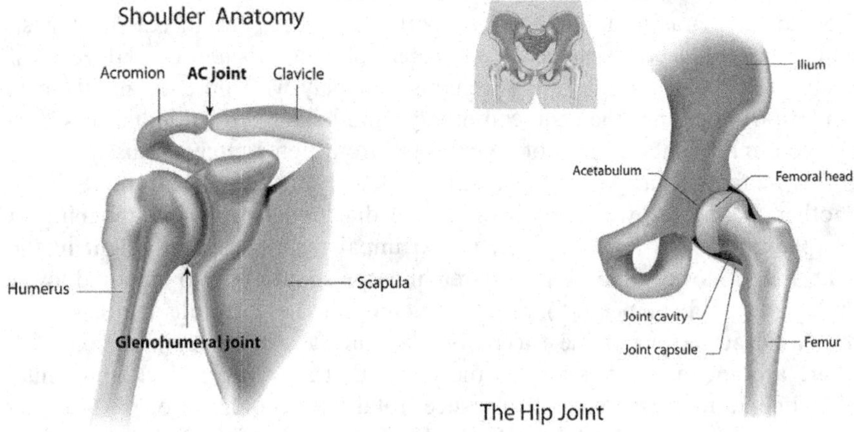

Figure 4.7 Skeletal restrictions on ROM. Hip acetabulum has a much deeper cup-like structure versus the flatter more mobile shoulder glenoid fossa

with the glenoid fossa (see Figure 4.7). It can also perform flexion/extension, abduction/adduction, medial and lateral rotation and circumduction. Whereas, shoulder motion is expansive, hip ROM is quite limited compared with the shoulder.

Normal ROM (degrees)	*Hip*	*Shoulder*
Flexion	110–120	180
Extension	10–15	45–60
Abduction	30–50	150
External rotation (lateral)	40–60	90
Internal rotation (medial)	30–40	70–90

Thus, skeletal structures are typically not modifiable with any amount or type of warm-up and stretching will typically not affect the skeletal structures.

Ligaments

Ligaments are tough, fibrous tissue that connect one bone to another bone providing joint stability. An important function is to prevent movement that might damage a joint. In most circumstances, the objective of stretching is NOT to elongate the ligaments. Lengthening ligaments would decrease joint stability and often lead to injury. There are some cases however, when the goal of stretching would be to elongate ligaments. Athletes who need to attain extreme levels of flexibility such as gymnasts, figure skaters, dancers, certain circus performers and others may target their ligaments.

Some individuals have a predisposition to hyper-laxity of their connective tissue or hypermobility of their joints. Women tend to have greater joint laxity or hypermobility compared with men (16, 18, 33, 63, 120). Harry Houdini (1874–1926) was famous for his escape abilities (see Figure 4.8). He could be tied up in ropes or chains and submerged in water or other hazards (i.e., suspended from buildings) and would make miraculous escapes. Houdini had lax ligaments and could subluxate (partially dislocate) or dislocate many of his joints. Hence, movements that were impossible for the average individual, were possible for Houdini by voluntarily subluxating a joint, sliding or moving out of a chain or rope and then popping the joint back into place before reappearing before the audience. Naturally, you should not need to be warned that you should not try this at home. Average individuals who experience an injury with a joint dislocation or subluxation often have difficulties throughout their life with subsequent dislocations. People with hypermobile joints (see Figure 4.9) can experience increased chances for nerve compression disorders (99), impaired proprioception (10, 74) increased risk of joint trauma and osteoarthritis (17, 43, 49). Ligaments lack the extensive vascularity of muscles and tendons and tend not to heal or recover completely back to their original length and tension. So, unless there is a need for

Figure 4.8 Harry Houdini

Figure 4.9 Contortionist with hypermobile joints

extreme ROM capabilities by an athlete, stretching for health and normal function should not be so severe as to lengthen the ligaments (includes joint capsule). However, placing mechanical stress on ligaments as with stretching can play an important role in the processes of cellular differentiation up-regulating (increasing the activity of) ligament fibroblast markers such as collagen types I and III and tenascin-C (4). Thus stretch-induced mechanical stress would help build the ligament matrix making it stronger and more resistant to damage or injury. Similar effects are seen with tendon tissue, where a lack of mechanical stress can cause a progressive loss of cell matrix and a prolonged period of stress deprivation results in even higher mechanical stresses needed to promote tendon tissue growth than before the removal of stress strain (6).

Adiposity

Excessive fat can affect flexibility tests. An abundance of adiposity around the waist and trunk will impede hip flexion for example. But there is no concrete evidence that fat will impede the extensibility of tissues. An early study by (114), did not find any significant correlation between low fat, high fat and muscular individuals and their flexibility. Alter (3) points out that very large sumo wrestlers with an overabundance of fat can still exhibit extraordinary levels of flexibility. Of course, fat is not a tissue targeted during stretching during a warm-up. Proper diet and exercise are the key to improving fat-induced restrictions on some ROM tests (i.e., sit and reach or toe touch).

Nerves

When stretching, can we elongate the nerves? Nerves do have the capacity to elongate to a certain point. The extensibility of the nerve resides mostly with the perineurium (connective tissue sheath surrounding a bundle (fascicle) of nerve fibres within a nerve) with an elongation range between 6–20 percent of its resting length (3, 107, 108). Once the maximum length is reached, the nerve is susceptible to tearing or injury. If the perineurium or nerve is ruptured, there can be leakage of proteins into the fascicles and edema and reduced possibility for regeneration (50). Furthermore, stretching the nerve as much as 8 percent can reduce blood flow with complete occlusion at 15 percent elongation (67, 68, 93). Nerve conduction can also be inhibited with only 6 percent elongation (50, 115). Stretching to induce neural tension (four x 1-minute stretches of the hamstrings with head, neck and trunk slumped down (partially flexed)) contributed to knee flexion strength loss (84). As long as the maximum elongation limit is not exceeded then the nerve should be able to return to its original length (109). With muscle fascicles able to elongate approximately 50 percent, how is it possible that we can stretch without always damaging our nerves? Although the nerve itself can only elongate 6–20 percent, the nerves are not typically situated in a straight line but rather are in a slacker position due to an undulating path through the tissues (fasciculi) (104). So, when stretching, the nerve does not initially elongate, but it actually straightens out till there is no longer an undulating course. Thus, the we can stretch far greater lengths than just 20 percent of the resting length of the musculotendinous tissues without damaging the nerves. Another evolutionary safety aspect is that most nerves traverse the flexor side of the joint. When a joint is flexed, the nerves would actually be placed in a more relaxed position rather than under the stress of an elongated position. Exceptions to this general rule are the ulnar nerve which traverse the extensor aspect of the elbow. You will experience the anatomical position of your ulnar nerve when you hit your "funny bone" and feel the "humour" of the pain running up your arm, as you hit the ulnar nerve near the humerus.

The sciatic nerve is another exception to the flexor positioning of nerves, as it runs across the extensor aspect of the hip. The sciatic nerve is protected by an especially thick epineurium, which constitutes about 88 percent of its cross-sectional area (108), as we spend an inordinate amount of time squatting (elongating the nerve tract) and sitting on the sciatic nerve (3). Although our primary focus is not stretching of the nerves, it has been shown that acute maximal dorsiflexion ROM can be increased by stretching the sciatic nerve without altering the muscle stiffness (5). The decrease in sciatic nerve stiffness was significantly correlated with the increase in maximal dorsiflexion ROM ($r = 0.571$). Therefore, while it is not a primary aim of stretching to elongate our nerves and associated connective sheaths, there can be a minor contribution to increasing ROM.

Muscle Hypertrophy

In the past, strength training resulting in muscle hypertrophy was feared to result in "muscle boundness". Being "muscle bound" inferred having large muscles that would inhibit flexibility. Contrary to this myth, there are examples of hypertrophied bodybuilders such as "Flex Wheeler" who could perform the Russian splits (legs spread laterally or fully abducted) during competitions. There are many other hypertrophied athletes such as American football players, rugby, ice hockey players, sumo wrestlers, and others who exhibit superb levels of flexibility. Although performing partial ROM (also known as cheat "reps") resistance training can shorten a muscle inhibiting ROM, there are a number of studies (66, 78, 101, 118, 119) and a recent meta-analysis (2) demonstrating increased ROM when performing full ROM resistance training These resistance training ROM improvements are also seen with senior adults (9, 40). Although not every study finds an increase in flexibility with resistance training, there is typically no loss of ROM (48). Thus, large muscles have a great propensity for extensibility.

Intracellular and Extracellular Connective Tissue

Muscles have both intracellular and extracellular components. When stretching a muscle, what parts are actually being stretched? Myofibrils can be stretched to twice their resting length without damage (47). The elasticity of the myofibrils is largely due to an intracellular protein called titin, with a length of approximately half the sarcomere being anchored at the Z disc (through a telethonin protein) and the M line (see Figure 4.10) (47). Titin can act as a spring that unfolds in response to high tension or stress. Myofibrils are connected together by the extracellular matrix. An essential protein in the extracellular matrix is fibronectin, which has elastic properties. Fibronectin can stretch up to four times its resting relaxed length. Fibronectin is composed of three subunits (FN-I, FN-II, and FM-III) with FN-III as the subunit that unfolds to contribute to the elasticity (see Figure 4.11). Stressing these units either mechanically, thermally or chemically leads to further binding with other fragments (e.g., anastellin) to form superfibronectins with increased adhesion capabilities. Fibronectins are also in contact with integrin molecules on the muscle membrane (47). Integrin molecules also sense membrane tension and transfer this information to the Raptor-mTOR complex in the nucleus helping to promote greater protein synthesis (58, 123). Specifically, integrin links laminin in the extracellular matrix with the cell cytoskeleton and translates mechanical forces into chemical signals. Stretching and muscle contractions activate intracellular signaling molecules that respond to injury-induced damage. Integrin stabilizes the muscle and provides communication between the matrix and cytoskeleton. Thus, elevated tension from stretching detected by the integrin molecule transduced to the nucleus leads to increased protein synthesis and helps with injury protection and prevention (21).

This enhanced muscle protein synthesis would increase muscle volume and strength, protecting it from other similar force stressors. An animal experiment

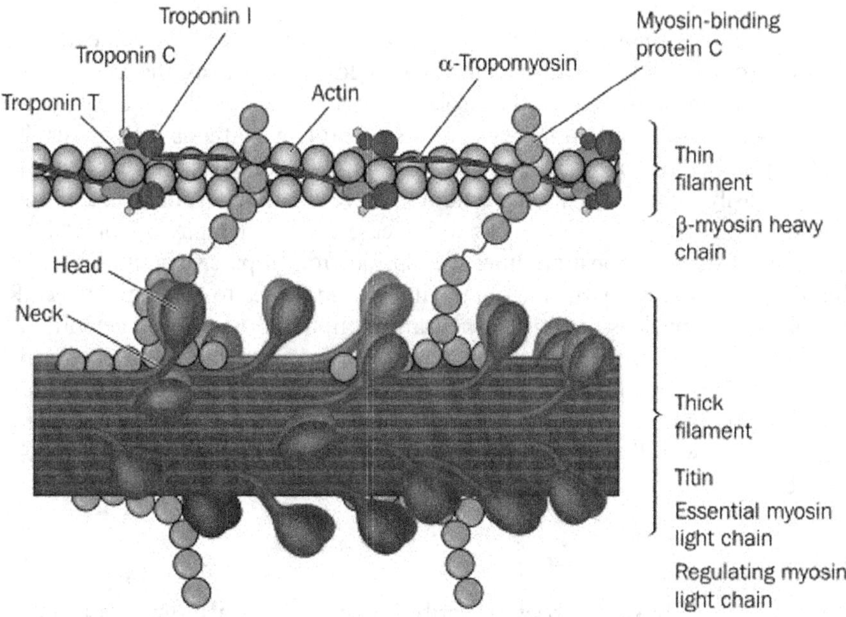

Figure 4.10 Titin and myofilament structure

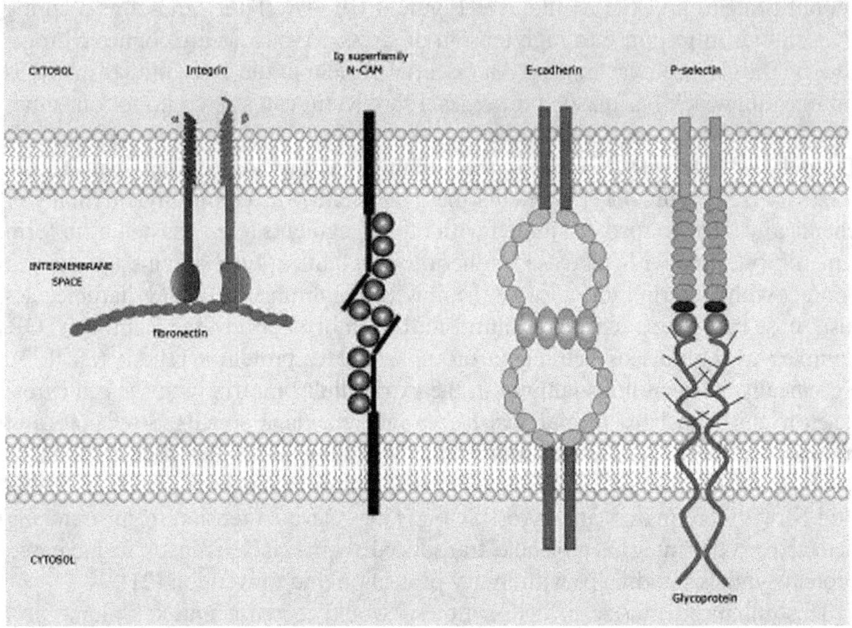

Figure 4.11 Fibronectin and integrin

had chicken latissimus dorsi (lats) muscles passively stretched over an extended duration (days). Fortunately, or unfortunately for the chickens, they were sacrificed and the muscles were used in isolation in a physiological saline bath solution. Researchers found an increase in muscle protein synthesis activity just by passively stretching the isolated muscle (64). Not all animal model experiments transfer easily to humans, and I would not suggest hanging from a bar or being placed on a rack for days or weeks to increase the hypertrophy of your "lats". However, it does demonstrate how the tension from a passive stretch is transduced (signal is sent) by the membrane integrin molecule to the Raptor-mTOR complex in the nucleus to activate the transcription and translation activities of the DNA and DNA respectively to promote muscle growth (increased protein synthesis). This stretch-induced adaptation may contribute to the increased tolerance of the muscle matrix to external forces and torques decreasing injury incidence. We will discuss in-depth in another chapter, the evidence for stretch-induced increases in muscle strength and hypertrophy.

Tendons

Static stretching has often been prescribed as part of a warm-up or a training routine, as it is purported to decrease the incidence of musculotendinous injuries (12, 83). The rationale for decreased musculotendinous injuries would be that the muscles and tendons would be better able to withstand or absorb the forces placed on the tissues and not tear (muscle strain or tendon sprain). Logically, then, increased ROM should not only be ascribed to decreased viscosity, neural inhibition, and stretch tolerance but also decreased stiffness or increased compliance of the muscles and tendons. There is much conflict in the literature as to the extent of tissue compliance/stiffness changes. Some researchers claim an association between reductions in myotendinous stiffness and ROM following an acute bout of static stretching (89), suggesting altered muscle mechanical properties are important contributors (89). Two very important stretch researchers, Anthony Blazevich (Australia) and Anthony Kay (UK) investigated the contribution of muscle and tendon elongation to maximum dorsiflexor's ROM. They found that muscular tissue is more compliant than the tendon during passive stretch. In their study, muscle lengthened 14.9 percent compared with 8.4 percent lengthening of the tendon under maximal stretch conditions (19). Thus, the maximum dorsiflexion ankle ROM was not constrained by tendon elongation because further tendon strain should have been possible. Similarly, Abellaneda, (1) from the Guissard and Duchateau lab, indicated that the relative contribution of muscle fascicles to MTU elongation was 71.8 percent, whereas the tendon contributed 28.2 percent. Tendons are composed of 60–85 percent collagen protein (62), whereas connective tissue in muscle constitutes between 1 and 10 percent. Collagen is a very strong fibrous tissue explaining the decreased tendon compliance. It is important to note that these studies show the response of muscle and tendon to stretching. However, prior to stretching, the resting length of the tendon far exceeds the muscle fascicles and thus the tendon can account for about half of the change in musculotendinous unit length. Because tendon elongation exceeds muscle, then stress-induced muscle strain surpasses tendon strain by about four fold (70). Freitas et al. (45) in their meta-analysis of 26 stretch training studies found that there were trivial effects of stretch training on tendon stiffness whereas one study actually demonstrated lower tendon extensibility after three weeks of stretch training (20). Hence, while tendons contribute greatly to ROM, the lesser elongation of the muscle fascicles causes them to receive more of the strain caused by stretching, thus leading to a greater chance for muscle strains versus tendon sprains.

Fascicles

Another means to increase muscle length might be to change both the length and the angle of the muscle fascicles (fibres). In that same study by Blazevich et al. (19), they reported that the contribution of changes in fascicle angle to the maximum stretch capabilities was negligible compared with the contributions of fascicle elongation. When comparing highly flexible gymnasts to volleyball players, rhythmic gymnasts displayed longer muscle fascicles at rest and greater fascicle elongation (distal part of medial gastrocnemius)

(35). There is also a minor contribution to increased muscle elongation from fascicle rotation. A significant difference between flexible and inflexible individuals is that fascicle rotation during stretch was greater in flexible subjects (~40 percent at 30° dorsiflexion) than inflexible subjects (~25 percent) (19). Thus, rotation permits interfascicle (and interfibre) translation. As there has been reported to be little difference between flexible and inflexible individuals in muscle lengthening and changes in fascicle length (series elastic component), the significant differences in fascicle rotation suggest a substantial difference in the parallel elastic component as well as the previously reported greater stretch tolerances of flexible individuals. The role of myofascia with muscle force and flexibility has become a much more involved debate. Deep fascia is also considered a key element to transmit load in parallel bypassing the joints, transmitting around 30 percent of the force generated during a muscular contraction.

Acute Morphological Proprioceptive Neuromuscular Facilitation (PNF) Stretching Mechanisms

Similar morphological mechanisms might be expected with PNF stretching. Static stretching is reported to be more effective in decreasing muscle stiffness (90), whereas PNF is more efficient in reducing both muscle and tendon stiffness (60). Perhaps, as PNF is reported to increase stretch tolerance more than static stretching (90), the higher stretch tolerance may provide greater stress upon both the muscle and the tendon. It might be rationalized that the contractions employed with contract-relax (CR) and contract-relax-agonist-contract (CRAC) PNF stretching provides a mental distraction (8), permitting a greater pain or stretch tolerance. As mentioned previously, the literature is conflicted as to whether PNF actually provides greater (39, 42, 46, 92), similar (28, 69) or trivial (85), increases in ROM compared with static stretching or even less flexibility than static stretching (111).

Acute Morphological Dynamic Stretching Mechanisms

As dynamic stretching uses repeated cyclical loading and unloading of the involved muscles, usually for a few minutes (44), these muscular contractions should induce shear stresses between muscle fibres and increase muscle temperature decreasing viscosity (thixotropy). In animal models, repeated lengthening and shortening of a muscle has increased muscle extensibility (88). There is one human dynamic stretching study indicating passive muscle stiffness reductions with increased ROM (55).

Dynamic, ballistic stretching has often not been advocated for efficiently improving ROM. Higher velocity dynamic stretching (ballistic) can weaken tissue (116, 117) by producing greater tensile forces over a brief duration (110). These high forces within a short time period do not allow for stress relaxation or creep to occur. Stress relaxation and creep refer to the reduction in tension and tissue lengthening, respectively, that occur when a tissue is held in a lengthened position for an extended period of time (3). Hence, these mechanisms may explain why ballistic stretching is often found not to provide substantial increases in ROM and the fear that ballistic stretching is more likely to lead to injuries especially with muscle that is not warmed up. Unfortunately, there is limited literature rationalizing the ROM morphological mechanisms for dynamic stretching.

Summary

Increased flexibility can be achieved even without stretching. Increasing tissue (muscles and tendons) temperatures elicit thixotropic effects, which decrease tissue visco-elasticity. Static and PNF stretching can activate a number of inhibitory reflexes. Prolonged static modes of stretching can disfacilitate spindle reflexes as well as inducing pre- and post-synaptic inhibition of the afferents. Specifically, the activity (discharge frequency) of intrafusal, muscle spindle, nuclear chain (detects extent of stretch) and nuclear bag (detects the extent and rate of stretch) fibres diminish with the static modes of stretching. Although GTO are predominately inhibitory, they do not respond strongly to stretch and any extent of inhibition from GTO subside almost immediately following the stretch. Furthermore, Renshaw cells also do not play a substantial inhibitory role with static stretching. However, extensive static stretching can activate cutaneous afferent inhibition. Active static and PNF stretching, which involve the contraction of the antagonist muscle, could initiate the contribution of reciprocal inhibition. PNF stretching also uses many of these inhibitory mechanisms, whereas dynamic stretching tends to excite rather than inhibit these reflexes. Hence, the stretch tolerance theory can also help explain increased ROM as the individual accommodates the discomfort or pain associated with stretching and can push themselves past their prior ROM.

Morphological considerations include the skeletal configuration and alignment which cannot be modified with stretching. Ligaments which help secure bone to bone (joint stability) are primarily inelastic and avascular and thus are also resistant to elastic or plastic elongation by stretching except with the intense flexibility training routines of extreme athletes like gymnasts, dancers, figure skaters and others. Excessive fat can impede joint ROM. Nerves can acutely elongate approximately 6–20 percent of their resting length but thereafter they are susceptible to injury. Highly hypertrophied muscle can also provide some ROM restrictions, however the great extensibility of muscle puts into dispute the old theory of muscle boundness. Myofibrils can elongate double their resting length mainly due to the protein titin. Muscle is more compliant than tendons, with tendons accounting for less than half of the musculotendinous unit change. Muscle extensibility can also be altered by changes in fascicle angle (trivial), fascicle rotation (minor), and fascicle elongation (substantial). However, stretch tolerance may provide a greater impetus to enhanced ROM than decreased musculotendinous stiffness.

References

1. Abellaneda, S., Guissard, N., and Duchateau, J.The relative lengthening of the myotendinous structures in the medial gastrocnemius during passive stretching differs among individuals. *J Appl Physiol* 106: 169–177, 2009.
2. Alizadeh, S., Daneshjoo, A., Zahiri, A., Anvar, S.H., Goudini, R., Hicks, J.P., Konrad, A., and Behm, D.G.Resistance Training Induces Improvements in Range of Motion: A Systematic Review and Meta-Analysis. *Sports Med* 53: 707–722, 2023.
3. Alter, M.J. *Science of Flexibility.* Champaign, IL: Human Kinetics, 1996.
4. Altman, G.H., Horan, R.L., Martin, I., Farhadi, J., Stark, P.R., Volloch, V., Richmond, J.C., Vunjak-Novakovic, G., and Kaplan, D.L.Cell differentiation by mechanical stress. *FASEB J* 16: 270–272, 2002.
5. Andrade, R.J., Freitas, S.R., Hug, F., Le Sant, G., Lacourpaille, L., Gross, R., McNair, P., and Nordez, A.The potential role of sciatic nerve stiffness in the limitation of maximal ankle range of motion. *Sci Rep* 8: 14532, 2018.
6. Arnoczky, S.P., Lavagnino, M., Egerbacher, M., Caballero, O., Gardner, K., and Shender, M.A.Loss of homeostatic strain alters mechanostat "set point" of tendon cells in vitro. *Clin Orthop Relat Res* 466: 1583–1591, 2008.
7. Avela, J., Kyröläinen, H., and Komi, P.V.Altered reflex sensitivity after repeated and prolonged passive muscle stretching. *Journal of Applied Physiology* 86: 1283–1291, 1999.
8. Azevedo, D.C., Melo, R.M., Alves Correa, R.V., and Chalmers, G.Uninvolved versus target muscle contraction during contract: relax proprioceptive neuromuscular facilitation stretching. *Phys Ther Sport* 12: 117–121, 2011.
9. Barbosa, A.R., Santarem, J.M., Filho, W.J., and Marucci Mde, F.Effects of resistance training on the sit-and-reach test in elderly women. *J Strength Cond Res* 16: 14–18, 2002.
10. Barrack, R.L., Skinner, H.B., Brunet, M.E., and Cook, S.D.Joint Laxity and Proprioception in the Knee. *Phys Sportsmed* 11: 130–135, 1983.
11. Behm, D.G., Bambury, A., Cahill, F., and Power, K.Effect of acute static stretching on force, balance, reaction time, and movement time. *Med Sci Sports Exerc* 36: 1397–1402, 2004.
12. Behm, D.G., Blazevich, A.J., Kay, A.D., and McHugh, M.Acute effects of muscle stretching on physical performance, range of motion, and injury incidence in healthy active individuals: a systematic review. *Appl Physiol Nutr Metab* 41: 1–11, 2016.
13. Behm, D.G., Button, D.C., and Butt, J.C.Factors affecting force loss with prolonged stretching. *Can J Appl Physiol* 26: 261–272, 2001.
14. Behm, D.G., Cavanaugh, T., Quigley, P., Reid, J.C., Nardi, P.S., and Marchetti, P. H.Acute bouts of upper and lower body static and dynamic stretching increase non-local joint range of motion. *Eur J Appl Physiol* 116: 241–249, 2016.
15. Behm, D.G. and Chaouachi, A.A review of the acute effects of static and dynamic stretching on performance. *Eur J Appl Physiol* 111: 2633–2651, 2011.
16. Beighton, P., Solomon, L., and Soskolne, C.L.Articular mobility in an African population. *Ann Rheum Dis* 32: 413–418, 1973.
17. Beighton, P., Grahame, R., and Bird, H. *Hypermobility of joints.* London: Springer-Verlag, 1999.
18. Biro, F., Gewanter, H.L., and Baum, J.The hypermobility syndrome. *Pediatrics* 72: 701–706, 1983.
19. Blazevich, A.J., Cannavan, D., Waugh, C.M., Fath, F., Miller, S.C., and Kay, A.D. Neuromuscular factors influencing the maximum stretch limit of the human plantar flexors. *Journal of Applied Physiology* 113: 1446–1455, 2012.

20. Blazevich, A.J., Cannavan, D., Waugh, C.M., Miller, S.C., Thorlund, J.B., Aagaard, P., and Kay, A.D.Range of motion, neuromechanical, and architectural adaptations to plantar flexor stretch training in humans. *J Appl Physiol (1985)* 117: 452–462, 2014.
21. Boppart, M.D., Burkin, D.J., and Kaufman, S.J.Alpha7beta1-integrin regulates mechanotransduction and prevents skeletal muscle injury. *Am J Physiol Cell Physiol* 290: C1660–1665, 2006.
22. Budini, F., Christova, M., Gallasch, E., Kressnik, P., Rafolt, D., and Tilp, M. Transient Increase in Cortical Excitability Following Static Stretching of Plantar Flexor Muscles. *Front Physiol* 9: 530, 2018.
23. Budini, F., Christova, M., Gallasch, E., Rafolt, D., and Tilp, M.Soleus H-Reflex Inhibition Decreases During 30 s Static Stretching of Plantar Flexors, Showing Two Recovery Steps. *Front Physiol* 9: 935, 2018.
24. Budini, F., Rafolt, D., Christova, M., Gallasch, E., and Tilp, M.The Recovery of Muscle Spindle Sensitivity Following Stretching Is Promoted by Isometric but Not by Dynamic Muscle Contractions. *Front Physiol* 11: 905, 2020.
25. Bussel, B. and Pierrot-Deseilligny, E.Inhibition of human motoneurons, probably of Renshaw origin, elicited by an orthodromic motor discharge. *J Physiol* 269: 319–339, 1977.
26. Chalmers, G.Re-examination of the possible role of Golgi tendon organ and muscle spindle reflexes in proprioceptive neuromuscular facilitation muscle stretching. *Sports Biomech* 3: 159–183, 2004.
27. Chaouachi, A., Padulo, J., Kasmi, S., Othmen, A.B., Chatra, M., and Behm, D.G. Unilateral static and dynamic hamstrings stretching increases contralateral hip flexion range of motion. *Clinical Physiology and Functional Imaging*, 2015.
28. Condon, S.M. and Hutton, R.S.Soleus muscle electromyographic activity and ankle dorsiflexion range of motion during four stretching procedures. *Physical Therapy* 67: 24–30, 1987.
29. Crone, C. and Nielson, J.Spinal mechanisms in man contributing to reciprocal inhibition during voluntary dorsiflexion of the foot. *J Physiol (Lond)* 116: 255–272, 1989.
30. da Silva, J.J., Behm, D.G., Gomes, W.A., Silva, F.H., Soares, E.G., Serpa, E.P., Vilela Junior Gde, B., Lopes, C.R., and Marchetti, P.H.Unilateral plantar flexors static-stretching effects on ipsilateral and contralateral jump measures. *J Sports Sci Med* 14: 315–321, 2015.
31. Datoussaid, M., El Khalouqi, H., Dahm, C., Guissard, N., and Baudry, S.Passive torque influences the Hoffmann reflex pathway during the loading and unloading phases of plantar flexor muscles stretching. *Physiol Rep* 9: e14834, 2021.
32. de Oliveira, U.F., de Araujo, L.C., de Andrade, P.R., Dos Santos, H.H., Moreira, D.G., Sillero-Quintana, M., and de Almeida Ferreira, J.J.Skin temperature changes during muscular static stretching exercise. *J Exerc Rehabil* 14: 451–459, 2018.
33. Decoster, L.C., Vailas, J.C., Lindsay, R.H., and Williams, G.R.Prevalence and features of joint hypermobility among adolescent athletes. *Arch Pediatr Adolesc Med* 151: 989–992, 1997.
34. Delwaide, P.J., Toulouse, P., and Crenna, P.Hypothetical role of long-loop reflex pathways. *Appl Neurophysiol* 44: 171–176, 1981.
35. Donti, O., Panidis, I., Terzis, G., and Bogdanis, G.C.Gastrocnemius Medialis Architectural Properties at Rest and During Stretching in Female Athletes with Different Flexibility Training Background. *Sports (Basel)* 7, 2019.

36. Drinkwater, E.J. and Behm, D.G.Effects of 22 degrees C muscle temperature on voluntary and evoked muscle properties during and after high-intensity exercise. *Appl Physiol Nutr Metab* 32: 1043–1051, 2007.
37. Edin, B.B. and Vallbo, A.B.Classification of human muscle stretch receptor afferents: a Bayesian approach. *J Neurophysiol* 63: 1314–1322, 1990.
38. Enoka, R.M., Hutton, R.S., and Eldred, E.Changes in excitability of tendon tap and Hoffmann reflexes following voluntary contractions. *Electroencephalogr Clin Neurophysiol* 48: 664–672, 1980.
39. Etnyre, B.R. and Lee, E.J.Chronic and Acute Flexibility of Men and Women Using 3 Different Stretching Techniques. *Research Quarterly for Exercise and Sport* 59: 222–228, 1988.
40. Fatouros, I.G., Taxildaris, K., Tokmakidis, S.P., Kalapotharakos, V., Aggelousis, N., Athanasopoulos, S., Zeeris, I., and Katrabasas, I.The effects of strength training, cardiovascular training and their combination on flexibility of inactive older adults. *Int J Sports Med* 23: 112–119, 2002.
41. Faulkner, J.A., Zerba, E., and Brooks, S.V.Muscle temperature of mammals: cooling impairs most functional properties. *Am J Physiol* 259: R259−R265, 1990.
42. Ferber, R., Osternig, L., and Gravelle, D.Effect of PNF stretch techniques on knee flexor muscle EMG activity in older adults. *J Electromyogr Kinesiol* 12: 391–397, 2002.
43. Finsterbush, A. and Pogrund, H.The hypermobility syndrome. Musculoskeletal complaints in 100 consecutive cases of generalized joint hypermobility. *Clin Orthop Relat Res*: 124–127, 1982.
44. Fletcher, I.M.The effect of different dynamic stretch velocities on jump performance. *Eur J Appl Physiol* 109: 491–498, 2010.
45. Freitas, S.R., Mendes, B., Le Sant, G., Andrade, R.J., Nordez, A., and Milanovic, Z.Can chronic stretching change the muscle-tendon mechanical properties? A review. *Scand J Med Sci Sports*, 2017.
46. Freitas, S.R., Vilarinho, D., Rocha Vaz, J., Bruno, P.M., Costa, P.B., and Milhomens, P.Responses to static stretching are dependent on stretch intensity and duration. *Clin Physiol Funct Imaging* 35: 478–484, 2015.
47. Gao, M., Sotomayor, M., Villa, E., Lee, E.H., and Schulten, K.Molecular mechanisms of cellular mechanics. *Phys Chem Chem Phys* 8: 3692–3706, 2006.
48. Girouard, C.K. and Hurley, B.F.Does strength training inhibit gains in range of motion from flexibility training in older adults? *Med Sci Sports Exerc* 27: 1444–1449, 1995.
49. Grahame, R.Joint hypermobility − clinical aspects. *Proc R Soc Med 64*: 692–694, 1971.
50. Grewal, R., Xu, J., Sotereanos, D.G., and Woo, S.L.Biomechanical properties of peripheral nerves. *Hand Clin* 12: 195–204, 1996.
51. Guissard, N. and Duchateau, J.Neural aspects of muscle stretching. *Exerc Sport Sci Rev* 34: 154–158, 2006.
52. Guissard, N., Duchateau, J., and Hainaut, K.Muscle stretching and motoneuron excitability. *European Journal of Applied Physiology* 58: 47–52, 1988.
53. Guissard, N., Duchateau, J., and Hainaut, K.Mechanisms of decreased motoneurone excitation during passive muscle stretching. *Experimental Brain Research* 137: 163–169, 2001.
54. Hanten, W.P., Olson, S.L., Butts, N.L., and Nowicki, A.L.Effectiveness of a home program of ischemic pressure followed by sustained stretch for treatment of myofascial trigger points. *Phys Ther* 80: 997–1003, 2000.

55. Herda, T.J., Herda, N.D., Costa, P.B., Walter-Herda, A.A., Valdez, A.M., and Cramer, J.T.The effects of dynamic stretching on the passive properties of the muscle-tendon unit. *J Sports Sci* 31: 479–487, 2013.
56. Hindle, K.B., Whitcomb, T.J., Briggs, W.O., and Hong, J.Proprioceptive neuromuscular facilitation (PNF): Its mechanisms and effects on range of motion and muscular function. *Journal of Human Kinetics* 31: 105–113, 2012.
57. Houk, J.C., Crago, P.E., and Rymer, W.Z.Functional properties of the Golgi tendon organs. In: *Spinal and Supraspinal Mechanisms of Voluntary Motor Control and Locomotion*. J.E. Desmedt, ed., 1980, pp. 33–43.
58. Huang, H., Kamm, R.D., and Lee, R.T.Cell mechanics and mechanotransduction: pathways, probes, and physiology. *Am J Physiol Cell Physiol* 287: C1–11, 2004.
59. Katz, R. and Pierrot-Deseilligny, E.Recurrent inhibition in humans. *Prog Neurobiol* 57: 325–355, 1999.
60. Kay, A.D., Husbands-Beasley, J., and Blazevich, A.J.Effects of Contract-Relax, Static Stretching, and Isometric Contractions on Muscle-Tendon Mechanics. *Med Sci Sports Exerc* 47: 2181–2190, 2015.
61. Khan, S.I. and Burne, J.A.Afferents contributing to autogenic inhibition of gastrocnemius following electrical stimulation of its tendon. *Brain Res* 1282: 28–37, 2009.
62. Kjaer, M.Role of extracellular matrix in adaptation of tendon and skeletal muscle to mechanical loading. *Physiol Rev* 84: 649–698, 2004.
63. Larsson, L.G., Baum, J., and Mudholkar, G.S.Hypermobility: features and differential incidence between the sexes. *Arthritis Rheum* 30: 1426–1430, 1987.
64. Laurent, G. and Sparrow, M.Changes in RNA, DNA and protein content and the rates of protein synthesis and degradation during hypertrophy of the anterior latissimus dorsi muscle of the adult fowl (gallus domesticus). *Growth* 41: 249–262, 1977.
65. Le Bars, D., Villanueva, L., Bouhassira, D., and Willer, J.C.Diffuse noxious inhibitory controls (DNIC) in animals and in man. *Patol Fiziol Eksp Ter*: 55–65, 1992.
66. Leighton, J.R.Flexibility characteristics of males ten to eighteen years of age. *Arch Phys Med Rehabil* 37: 494–499, 1956.
67. Lundborg, G.Structure and function of the intraneural microvessels as related to trauma, edema formation, and nerve function. *J Bone Joint Surg Am* 57: 938–948, 1975.
68. Lundborg, G. and Rydevik, B.Effects of stretching the tibial nerve of the rabbit. A preliminary study of the intraneural circulation and the barrier function of the perineurium. *J Bone Joint Surg Br* 55: 390–401, 1973.
69. Maddigan, M.E., Peach, A.A., and Behm, D.G.A comparison of assisted and unassisted proprioceptive neuromuscular facilitation techniques and static stretching. *Journal of strength and conditioning research/National Strength & Conditioning Association* 26: 1238–1244, 2012.
70. Magnusson, S.P., Narici, M.V., Maganaris, C.N., and Kjaer, M.Human tendon behaviour and adaptation, in vivo. *J Physiol* 586: 71–81, 2008.
71. Magnusson, S.P., Simonsen, E.B., Aagaard, P., Dyhre-Poulsen, P., McHugh, M.P., and Kjaer, M.Mechanical and physical responses to stretching with and without preisometric contraction in human skeletal muscle. *Arch Phys Med Rehabil* 77: 373–378, 1996.
72. Magnusson, S.P., Simonsen, E.B., Aagaard, P., Sorensen, H., and Kjaer, M.A mechanism for altered flexibility in human skeletal muscle. *Journal of Physiology* 497 (Pt 1): 291–298, 1996.

73. Magnusson, S.P., Simonsen, E.B., Dyhre-Poulsen, P., Aagaard, P., Mohr, T., and Kjaer, M.Viscoelastic stress relaxation during static stretch in human skeletal muscle in the absence of EMG activity. *Scand J Med Sci Sports* 6: 323–328, 1996.
74. Mallik, A.K., Ferrell, W.R., McDonald, A.G., and Sturrock, R.D.Impaired proprioceptive acuity at the proximal interphalangeal joint in patients with the hypermobility syndrome. *Br J Rheumatol* 33: 631–637, 1994.
75. Marchetti, P.H., Silva, F.H., Soares, E.G., Serpa, E.P., Nardi, P.S., Vilela Gde, B., and Behm, D.G.Upper limb static-stretching protocol decreases maximal concentric jump performance. *J Sports Sci Med* 13: 945–950, 2014.
76. Marchetti, P.H., Reis, R.G., Gomes, W.A., da Silva, W.A., Soares, E.G., de Freitas, F.S., and Behm, D.G.Static-stretching of the pectoralis major decreases tríceps brachii activation during a maximal isometric bench press. *Gazzetta Medica Italiana*, 2017.
77. Marchettini, P.H.Muscle pain: animal and human experimental and clinical studies. *Muscle Nerve* 16: 1033–1039, 1993.
78. Massey, B.A. and Chaudet, N.L.Effects of systematic, heavy resistance exercise on range of joint movements in young adults. *Research Quarterly* 27: 41–51, 1956.
79. Mattei, B., Schmied, A., Mazzocchio, R., Decchi, B., Rossi, A., and Vedel, J.P. Pharmacologically induced enhancement of recurrent inhibition in humans: effects on motoneurone discharge patterns. *J Physiol* 548: 615–629, 2003.
80. Matthews, P.B.C.Developing views on the muscle spindle. In: *Spinal and Supraspinal Mechanisms of Voluntary Motor Control and Locomotion*. 1980, pp. 12–27.
81. Matthews, P.B.C.Muscle spindles: their messages and their fusimotor supply. In: *The Nervous System: Handbook Of Physiology.* V.B. Brooks, ed.: American Physiological Society, 1981, pp. 189–288.
82. McArdle, W.D., Katch, F.I., and Katch, V.L. *Exercise Physiology: Energy, Nutrition, and Human Performance.* Malvern, PA: Lea and Febiger, 1991.
83. McHugh, M.P. and Cosgrave, C.H.To stretch or not to stretch: the role of stretching in injury prevention and performance. *Scand J Med Sci Sports* 20: 169–181, 2010.
84. McHugh, M.P., Johnson, C.D., and Morrison, R.H.The role of neural tension in hamstring flexibility. *Scand J Med Sci Sports* 22: 164–169, 2012.
85. Medeiros, D.M.M.D.Comparison between static stretching and proprioceptive neuromuscular facilitation on hamstring flexibility: Systematic review and meta-analysis. *European Journal of Physiotherapy*, 2018.
86. Mitchell, U.H., Myrer, J.W., Hopkins, J.T., Hunter, I., Feland, J.B., and Hilton, S. C.Acute stretch perception alteration contributes to the success of the PNF "contract-relax" stretch. *J Sport Rehabil* 16: 85–92, 2007.
87. Mitchell, U.H., Myrer, J.W., Hopkins, J.T., Hunter, I., Feland, J.B., and Hilton, S. C.Neurophysiological reflex mechanisms' lack of contribution to the success of PNF stretches. *J Sport Rehabil* 18: 343–357, 2009.
88. Mutungi, G. and Ranatunga, K.W.Temperature-dependent changes in the viscoelasticity of intact resting mammalian (rat) fast- and slow-twitch muscle fibres. *J Physiol* 508 (Pt 1): 253–265, 1998.
89. Nakamura, M., Ikezoe, T., Kobayashi, T., Umegaki, H., Takeno, Y., Nishishita, S., and Ichihashi, N.Acute effects of static stretching on muscle hardness of the medial gastrocnemius muscle belly in humans: an ultrasonic shear-wave elastography study. *Ultrasound Med Biol* 40: 1991–1997, 2014.
90. Nakamura, M., Ikezoe, T., Tokugawa, T., and Ichihashi, N.Acute Effects of Stretching on Passive Properties of Human Gastrocnemius Muscle-Tendon Unit:

Analysis of Differences Between Hold-Relax and Static Stretching. *J Sport Rehabil* 24: 286–292, 2015.

91. Nakamura M., Sato, S., Kiyono, R., Yahata, K., Yoshida, R., Fukaya, T., Nishishita, S., and Konrad, A.Association between the Range of Motion and Passive Property of the Gastrocnemius Muscle–Tendon Unit in Older Population. *Healthcare (Basel)* 9: 314–320, 2021.
92. O'Hora, J., Cartwright, A., Wade, C.D., Hough, A.D., and Shum, G.L.Efficacy of static stretching and proprioceptive neuromuscular facilitation stretch on hamstrings length after a single session. *Journal of Strength and Conditioning Research/National Strength & Conditioning Association* 25: 1586–1591, 2011.
93. Ogata, K. and Naito, M.Blood flow of peripheral nerve effects of dissection, stretching and compression. *J Hand Surg Br* 11: 10–14, 1986.
94. Opplert, J., Paizis, C., Papitsa, A., Blazevich, A.J., Cometti, C., and Babault, N. Static stretch and dynamic muscle activity induce acute similar increase in corticospinal excitability. *PLoS One* 15: e0230388, 2020.
95. Petersen, N., Morita, H., and Nielsen, J.Evaluation of reciprocal inhibition of the soleus H-reflex during tonic plantar flexion in man. *J Neurosci Methods* 84: 1–8, 1998.
96. Petersen, N., Morita, H., and Nielsen, J.Modulation of reciprocal inhibition between ankle extensors and flexors during walking in man. *Journal of Physiology* 520: 605–619, 1999.
97. Proske, U.What is the role of muscle receptors in proprioception? *Muscle & Nerve* 31: 780–787, 2005.
98. Pyndt, H., Laursen, M., and Nielsen, J.Changes in reciprocal inhibition across the ankle joint with changes in external load and pedaling rate during bicycling. *Journal Of Neurophysiology* 90: 3168–3177, 2003.
99. Russek, L.N.Hypermobility syndrome. *Phys Ther* 79: 591–599, 1999.
100. Sato, S., Kiyono, R., Takahashi, N., Yoshida, T., Takeuchi, K, and Nakamura, M.The acute and prolonged effects of 20-s static stretching on muscle strength and shear elastic modulus. *PLoS One* 15: e0228583, 2020.
101. Schmitt, G.D., Pelham, T.W., and Holt, L.E.Changes in flexibility of elite female soccer players resulting from a flexibility program or combined flexibility and strength program: A pilot study. *Clinical Kinesiology* 52: 64–67, 1998.
102. Shamos, M.H. and Lavine, L.S.Piezoelectricity as a fundamental property of biological tissues. *Nature* 213: 267–269, 1967.
103. Sharman, M.J., Cresswell, A.G., and Riek, S.Proprioceptive neuromuscular facilitation stretching: mechanisms and clinical implications. *Sports Med* 36: 929–939, 2006.
104. Smith, J.W.Factors influencing nerve repair. I. Blood supply of peripheral nerves. *Archives of Surgery* 93: 335–341, 1966.
105. Smith, N.P., Barclay, C.J., and Loiselle, D.S.The efficiency of muscle contraction. *Prog Biophys Mol Biol* 88: 1–58, 2005.
106. Stove, M.P., Hirata, R.P., and Palsson, T.S.Muscle stretching – the potential role of endogenous pain inhibitory modulation on stretch tolerance. *Scand J Pain* 19: 415–422, 2019.
107. Sunderland, S. *Traumatized nerves, roots and ganglia. Musculoskeletal factors and neuropathological consequences.* New York: Plenum Press, 1978.
108. Sunderland, S. *Nerve injuries and their repair. A Critical Appraisal.* London: Churchill Livingstone, 1991.

109. Sunderland, S. and Bradley, K.C.Stress-strain phenomena in human spinal nerve roots. *Brain* 84: 102–119, 1961.
110. Taylor, D.C., Dalton, J.D., Jr., Seaber, A.V., and Garrett, W.E., Jr. Viscoelastic properties of muscle-tendon units. The biomechanical effects of stretching. *Am J Sports Med* 18: 300–309, 1990.
111. Thomas, E., Bianco, A., Paoli, A., and Palma, A.The Relation Between Stretching Typology and Stretching Duration: The Effects on Range of Motion. *Int J Sports Med 39*: 243–254, 2018.
112. Timofeev, I., Grenier, F., and Steriade, M.Disfacilitation and active inhibition in the neocortex during the natural sleep-wake cycle: an intracellular study. *Proc Natl Acad Sci U S A* 98: 1924–1929, 2001.
113. Trajano, G.S., Nosaka, K., and Blazevich, A.J.Neurophysiological Mechanisms Underpinning Stretch-Induced Force Loss. *Sports Med*, 2017.
114. Tyrance, H.J.Relationships of extreme body types to ranges of flexibility. *Research Quarterly* 3: 349–359, 1958.
115. Wall, E.J., Massie, J.B., Kwan, M.K., Rydevik, B.L., Myers, R.R., and Garfin, S. R.Experimental stretch neuropathy. Changes in nerve conduction under tension. *J Bone Joint Surg Br* 74: 126–129, 1992.
116. Warren, C.G., Lehmann, J.F., and Koblanski, J.N.Elongation of rat tail tendon: effect of load and temperature. *Arch Phys Med Rehabil* 52: 465–474 passim, 1971.
117. Warren, C.G., Lehmann, J.F., and Koblanski, J.N.Heat and stretch procedures: an evaluation using rat tail tendon. *Arch Phys Med Rehabil* 57: 122–126, 1976.
118. Wickstrom, R.L.Weight training and flexibility. *Journal of Health, Physical Education and Recreation* 34: 61–62, 1963.
119. Wilmore, J.H., Parr, R.B., Girandola, R.N., Ward, P., Vodak, P.A., Barstow, T.J., Pipes, T.V., Romero, G.T., and Leslie, P.Physiological alterations consequent to circuit weight training. *Med Sci Sports* 10: 79–84, 1978.
120. Wordsworth, P., Ogilvie, D., Smith, R., and Sykes, B.Joint mobility with particular reference to racial variation and inherited connective tissue disorders. *Br J Rheumatol* 26: 9–12, 1987.
121. Young, W.B.The use of static stretching in warm-up for training and competition. *Int J Sports Physiol Perform* 2: 212–216, 2007.
122. Young, W.B. and Behm, D.G.Effects of running, static stretching and practice jumps on explosive force production and jumping performance. *J Sports Med Phys Fitness* 43: 21–27, 2003.
123. Zanchi, N.E. and Lancha, A.H., Jr. Mechanical stimuli of skeletal muscle: implications on mTOR/p70s6k and protein synthesis. *Eur J Appl Physiol* 102: 253–263, 2008.

5 Stretch Training-Related Range of Motion (ROM) Changes and Mechanisms

Chronic stretching (training) can increase range of motion (ROM) using different stretching techniques, positions, and durations (9). Naturally, there are always dissenting findings as for example; four (10 repetitions of 30 seconds; 3 days/week) (19) or six (four repetitions of 45 seconds; four days/week) (2) weeks of hamstrings static stretching did not increase hip extension or flexion ROM respectively. But the vast majority of stretch training programmes do exhibit ROM improvements (4, 7, 23, 26, 32). The most recent meta-analysis on the effects of chronic stretch training on ROM by the "king" of stretching and foam rolling meta-analyses, my friend and colleague Andreas Konrad reported that based on 77 studies that stretch training can provide moderate magnitude, effect size ROM increases (16).

But which type of stretch training is most effective? A review by Decoster et al. (9) indicated that static stretch training provided greater ROM improvements than proprioceptive neuromuscular facilitation (PNF). Other studies have documented static stretch-induced ROM improvements with no improvements with PNF (30 second stretch; three days/week for four weeks) (7), whereas another study found training-induced ROM improvements with no difference between static and PNF stretching (four days/week for six weeks) (32). While one study reported more than double the ROM improvements with static stretching versus dynamic stretching (1), others have not shown any significant difference (31). Once again, it is important to quantify the literature in general (meta-analysis) as reading individual studies can be quite confusing. Going back to the Konrad et al.'s meta-analysis (16), their subgroup analyses showed no difference in effectiveness between chronic static and PNF stretch training programmes, but both of these stretches produced greater ROM than ballistic and dynamic stretching. Furthermore, neither stretching volume, intensity, or weekly frequency demonstrated significant differences in ROM gains.

Their final significant finding was the evidence for greater training-related improvements in female ROM compared to males (16). This sex difference for stretch training contrasts with another meta-analysis (47 studies) of changes in ROM with an acute (single) bout of stretching, which did not find a sex difference (3). Thus, while women tend to possess greater absolute or inherent

DOI: 10.4324/9781032709086-6

flexibility than men and when performing chronic stretch training can achieve relatively greater ROM gains, they do not show such superiority with the relative flexibility increases following a single, acute session of stretching. In our acute meta-analysis (3), we suggested that although baseline flexibility is more limited in young or older adult males compared to females, the potential for acute ROM increases is not hindered by age, sex, or trained state. For the chronic stretch training results where women experienced greater relative ROM gains, we speculated that females do not exhibit higher ankle joint baseline flexibility compared to males. As the ankle joint was the most frequently tested joint in the literature, then the males would not have this explanation of a lower baseline from which to achieve higher relative increases. It was also postulated that women and girls may be more sensitive to chronic stretch training than men and boys.

The persistence of these flexibility adaptations after stretch-training desists have been reported to show better than pre-training ROM for three (20), four (29, 33), and eight (24) weeks. Similar to strength training, a maintenance flexibility training programme of one session per week can help preserve the training gains (30). Similar to elastic (acute) ROM responses, plastic ROM changes or changes attributed to flexibility training can also be attributed to a variety of factors including neural, morphological, and psychological. Which of these factors is most predominant is still under debate?

Plastic Neural Adaptations

Is it possible that the neural inhibitory responses that occur with elastic or acute changes in ROM can persist or be semi-permanent with training or plastic changes? Just as with the neural adaptations associated with strength training, prolonged alterations in the nervous system can evolve with flexibility training. Blazevich et al. (5) incorporated a three-week stretch training programme of the plantar flexors and found a reduction in tonic Ia (facilitatory) afferent feedback from muscle spindles (measured from the T-reflex). So, when the spindles were stretched, their firing frequency was less than before the flexibility training resulting in less reflex-induced contractions and a more relaxed muscle (disfacilitation). Perhaps, the decreased spindle activity was due to an increased compliance (reduced stiffness) of the passive elastic components. Thus, with a certain degree of stretch, the tissues around the spindles would accommodate the elongation and the spindle would not be stretched or elongated to the same extent resulting in a muted reflex response. However, in their study, the T-reflex reduction (spindle activity) did not parallel the reduction in passive stiffness during the stretch-training programme. Therefore, if the reduced spindle activity cannot be attributed to lower tissue compliance, then it is more likely that the flexibility training led to an intrinsic reduction in muscle spindle sensitivity. In this case, the stretch training had a direct effect on the activity of the nervous system (afferent input to the motoneurons). In the same Blazevich study, they also reported increased reciprocal inhibition in the soleus and gastrocnemius which would decrease the contractile force of the antagonists. Guissard and Duchateau had subjects stretch five days/week for six weeks (four stretch positions x five repetitions x 30 seconds each) and tested immediately after the training programme as well as after one month of detraining. Similar to the Blazevich et al. findings, they also found decreases in H- and T-reflexes (−36 percent and −14 percent respectively). However, they found that the reflex inhibition was present after the first ten stretching sessions but returned to baseline when tested after 30 days of detraining. However, passive muscle stiffness was apparent at every point from ten days of training to 30 days of detraining (15).

Plastic Morphological Adaptations

Stretch training studies in animal models have demonstrated sarcomerogenesis (increased number of sarcomeres in series) (8), but there is very limited or no evidence in humans. However, once again it should be pointed out that there are no longitudinal studies examining stretch-induced morphological changes over years of stretching. Again, with animal studies, stretch is a very potent signal for mechano growth factor (MGF: a variant of IGF-1), actin and myosin filament production, myosin isoform gene switching, protein turnover, hypertrophy as well as for adding sarcomeres in parallel and series (11, 12, 13, 14).

Traditionally PNF stretching effects were attributed primarily to neural inhibition (e.g., reciprocal inhibition, autogenic inhibition). However, there is a lack of experimental evidence for this emphasis on neural responses (6, 28). The effectiveness of PNF may be due to an increased tolerance to stretch (22) but there can also be morphological changes. For example, six weeks of PNF flexibility training reduced the passive and active stiffness of the Achilles tendon while also increasing the pennation angle of the gastrocnemius. However, there was no change in the passive resistive torque of the muscle (17). Even with 8–10-year old athletes, the flexibility trained athletes showed greater fascicle elongation when stretching with greater increases in joint angle and muscle-tendon unit displacement (27). Significant reductions in the myotendinous junction (muscle and tendon intersection) stiffness (e.g., 47 percent) during a passive static stretch have also been reported after stretch training (25) and this decreased passive stiffness can contribute to increased flexibility (15). On the contrary, a six week (five days/week) static stretch training programme improved hip flexion ROM but did not change muscle extensibility (4). The authors therefore attributed the improved ROM to increased stretch tolerance. Static stretch training (two days/week for 20 days) has been reported to reduce the tendon viscoelastic properties and but not the tendon stiffness (18) Static stretch training (seven days/week for six weeks) improved dorsiflexion ROM with a reduced muscle passive resistive torque but no change in tendon stiffness whereas ballistic stretch training exhibited the opposite effect with no change of muscle passive resistive torque but decreased tendon stiffness (23). A review on chronic stretching changes reported that stretch training durations of 3–8 weeks (average study: 5.1 weeks) do not alter muscle or tendon properties although it can increase the extensibility and tolerance of the muscle to a greater tensile force (10). They also reported that these durations of stretching also have trivial effects on tendon properties. While their review supports the stretch tolerance theory, you have to ask yourself if most athletes or fitness enthusiasts only stretch for five weeks or less than two months. Many athletes will have stretched on a regular basis from adolescence to early adulthood and may continue on a less vigorous or consistent basis for many years hence. Thus, examining studies of only a maximum of eight weeks duration does not provide a full picture of possible chronic morphological changes.

It would be fantastic for the public if the studies found consistent results, but each study may use different types of stretching, durations, volumes, intensities and subject populations. The most prudent comment or recommendation might be to incorporate a variety of stretching styles (static, dynamic, and PNF) in order to ensure that morphological changes to the muscle and tendons is optimized by providing a variety of stressors to the system.

Psychological Adaptations

While the training-induced increase in joint ROM can be connected to neural and morphological alterations, there is also a strong case for an increase in stretch tolerance (21, 22). This is also known as the sensory theory, which indicates the musculotendinous unit can tolerate greater tensions without a change in tension for a given length. Sixty individuals stretched for 30 minutes; five times per week for six weeks with no change in hamstrings' extensibility. As there was an increase in the ROM, the authors suggest that the improvement must have been due to greater stretch tolerance (4).

Summary

While there is still conflict in the literature, generally static and PNF stretching tend to provide greater improvements with static ROM than dynamic and ballistic stretching. Flexibility training adaptations can persist for 3–8 weeks with either reduced (once a week) or minimal stretching and activity. This persistent flexibility adaptations may be partially ascribed to neural adaptations such as an intrinsic disfacilitation of spindle afferent discharge. Although animal stretching studies have shown an increase in sarcomeres in series, there are no similar human findings with flexibility training. However, there is evidence with human chronic flexibility training for alterations in muscle pennation angles, visco-elastic properties and stretch tolerance.

References

1. Bandy, W.D., Irion, J.M., and Briggler, M.The effect of static stretch and dynamic range of motion training on the flexibility of the hamstring muscles. *J Orthop Sports PhysTher* 27: 295–300, 1998.
2. Bazett-Jones, D.M., Gibson, M.H., and McBride, J.M.Sprint and vertical jump performances are not affected by six weeks of static hamstring stretching. *J Strength Cond Res* 22: 25–31, 2008.
3. Behm, D.G., Alizadeh, S., Daneshjoo, A., Hadjizadeh Anvar, S., Graham, A., Zahiri, A., Goudini, R., Edwards, C., Culleton, R., Scharf, C., and Konrad, A. Acute effects of various stretching techniques on range of motion: A systematic review with meta-analysis. *Sports Medicine*, 2023.
4. Ben, M. and Harvey, L.A.Regular stretch does not increase muscle extensibility: a randomized controlled trial. *Scand J Med Sci Sports*, 2009.
5. Blazevich, A.J., Cannavan, D., Waugh, C.M., Fath, F., Miller, S.C., and Kay, A.D. Neuromuscular factors influencing the maximum stretch limit of the human plantar flexors. *Journal of Applied Physiology* 113: 1446–1455, 2012.
6. Chalmers, G.Re-examination of the possible role of Golgi tendon organ and muscle spindle reflexes in proprioceptive neuromuscular facilitation muscle stretching. *Sports Biomech* 3: 159–183, 2004.
7. Davis, D.S., Ashby, P.E., McHale, K.L., McQuain, J.A., and Wine, J.M.The effectiveness of 3 stretching techniques on hamstring flexibility using consistent stretching parameters. *Journal of Strength and Conditioning Research* 19: 27–32, 2005.
8. De Jaeger, D., Joumaa, V., and Herzog, W.Intermittent stretch training of rabbit plantarflexor muscles increases soleus mass and serial sarcomere number. *J Appl Physiol (1985)* 118: 1467–1473, 2015.
9. Decoster, L.C., Cleland, J., Altieri, C., and Russell, P.The effects of hamstring stretching on range of motion: a systematic literature review. *J Orthop Sports Phys Ther* 35: 377–387, 2005.
10. Freitas, S.R., Mendes, B., Le Sant, G., Andrade, R.J., Nordez, A., and Milanovic, Z.Can chronic stretching change the muscle-tendon mechanical properties? A review. *Scand J Med Sci Sports*, 2017.
12. Goldspink, D. F.The influence of immobilization and stretch on protein turnover of rat skeletal muscle. *Journal of Physiology* 264: 267–282, 1977.
11. Goldspink, D.F.The influence of passive stretch on the growth and protein turnover of the denervated extensor digitorum longus muscle. *Biochem J* 174: 595–602, 1978.
13. Goldspink, G.Changes in muscle mass and phenotype and the expression of autocrine and systemic growth factors by muscle in response to stretch and overload. *J Anat* 194 (Pt 3): 323–334, 1999.
14. Goldspink, G., Scutt, A., Martindale, J., Jaenicke, T., Gerlach, G., and Turay, L. Stretch and force generation induce rapid hypertrophy and myosin isoform gene switching in adult skeletal muscle. *Biochemical Society Transactions* 19: 368–373, 1990.
15. Guissard, N. and Duchateau, J.Effect of static stretch training on neural and mechanical properties of the human plantar-flexor muscles. *Muscle and Nerve* 29: 248–255, 2004.
16. Konrad, A., Alizadeh, S., Daneshjoo, A., Anvar, S.H., Graham, A., Zahiri, A., Goudini, R., Edwards, C., Scharf, C., and Behm, D.G.Chronic effects of stretching

on range of motion with consideration of potential moderating variables: A systematic review with meta-analysis. *J Sport Health Sci*, 2023.

17. Konrad, A., Gad, M., and Tilp, M.Effect of PNF stretching training on the properties of human muscle and tendon structures. *Scand J Med Sci Sports* 25: 346–355, 2015.
18. Kubo, K., Kanehisa, H., and Fukunaga, T.Effect of stretching training on the viscoelastic properties of human tendon structures in vivo. *Journal Applied Physiology* 92: 595–601, 2002.
19. LaRoche, D.P., Lussier, M.V., and Roy, S.J.Chronic stretching and voluntary muscle force. *J Strength Cond Res* 22: 589–596, 2008.
20. Long, P.A. *The effect of static, dynamic, and combined stretching exercise programs on the hip joint flexibility.* University of Maryland, 1971.
21. Magnusson, P. and Renstrom, P.The European College of Sports Sciences Position statement: The role of stretching exercises in sports. *European Journal of Sport Science* 6: 87–91, 2006.
22. Magnusson, S.P., Simonsen, E.B., Aagaard, P., Sorensen, H., and Kjaer, M.A mechanism for altered flexibility in human skeletal muscle. *Journal of Physiology* 497 (Pt 1): 291–298, 1996.
23. Mahieu, N.N., McNair, P., De Muynck, M., Stevens, V., Blanckaert, I., Smits, N., and Witvrouw, E.Effect of static and ballistic stretching on the muscle-tendon tissue properties. *MedSciSports Exerc* 39: 494–501, 2007.
24. McCue, B.F.Flexibility measurement of college women. *Research Quarterly* 3: 316–324, 1963.
25. Morse, C.I., Degens, H., Seynnes, O.R., Maganaris, C.N., and Jones, D.A.The acute effect of stretching on the passive stiffness of the human gastrocnemius muscle tendon unit. *J Physiol* 586: 97–106, 2008.
26. Nelson, A.G., Kokkonen, J., Eldredge, C., Cornwell, A., and Glickman-Weiss, E. Chronic stretching and running economy. *Scand J Med Sci Sports* 11: 260–265, 2001.
27. Panidi, I., Bogdanis, G.C., Gaspari, V., Spiliopoulou, P., Donti, A., Terzis, G., and Donti, O.Gastrocnemius Medialis Architectural Properties in Flexibility Trained and Not Trained Child Female Athletes: A Pilot Study. *Sports (Basel)* 8, 2020.
28. Sharman, M.J., Cresswell, A.G., and Riek, S.Proprioceptive neuromuscular facilitation stretching: mechanisms and clinical implications. *Sports Med* 36: 929–939, 2006.
29. Tweitmeyer, T.A. *A comparison of two stretching techniques for increasing and retaining flecibility.* University of Iowa, 1974.
30. Wallin, D., Ekblom, B., Grahn, R., and Nordenberg, T.Improvement of muscle flexibility: a comparison between two techniques. *The American Journal of Sports Medicine* 13: 263–267, 1985.
31. Wiemann, K. and Hahn, K.Influences of strength, stretching and circulatory exercises on flexibility parameters of the human hamstrings. *Int J Sports Med* 18: 340–346, 1997.
32. Yuktasir, B. and Kaya, F.Investigation into the long term effects of static and PNF stretching exercises on range of motion and jump performance. *Journal of Bodywork and Movement Therapies* 13: 11–21, 2009.
33. Zebas, C.J. and Rivera, M. L.Retention of flexibility in selected joints after cessation of a stretching exercise program. In: *Exercise physiology Current selected research*, Vol. I. New York: AMS Press, 1985, pp. 181–191.

6 Global Effects of Stretching

The process of science does not typically follow the amazing discoveries of Newton, Einstein, or Curie or our other historical luminaries. Even these amazing scientists had to build upon the prior work of others. Newton is reported to have stated, "If I have seen further, it is by standing on the shoulders of giants." These incredible scientists while building upon the foundational work of others, subsequently provided giant leaps over their predecessors. Most researchers are scientific bricklayers and we add a few more bricks to the foundation of knowledge and hopefully a bit like a good architect suggest slightly new designs (research directions) to investigate.

Crossover or non-local muscle fatigue

In the early 2000s a relatively new concept termed crossover fatigue was reported (36, 45). Surprisingly to many people, if you unilaterally fatigued one limb and then tested the same (homologous) contralateral muscle, impairments in muscle strength, power, or endurance could be experienced. I read those papers and was intrigued by the finding that one arm or leg muscle could produce less force or power and fatigue more quickly without even doing any physical activity. All you had to do was fatigue the opposite (contralateral) muscle. Naturally lacking any substantive imagination of my own, to come up with amazing new ideas, I borrowed their idea and found a research gap. The gap was that the early research typically examined the contralateral homologous muscle (unilaterally fatigue the quadriceps and test the contralateral quadriceps). Thus, my amazing revelation (sarcasm alert!) was to investigate if non-exercised contralateral or even ipsilateral (same side) heterologous muscles would be impaired by prior fatigue of another muscle. As it could be either contralateral or ipsilateral heterologous muscles, we thought crossover fatigue was not totally appropriate, so we coined the term "non-local muscle fatigue" (NLMF) as the impairments could be anywhere in the body (global effects) (21, 22, 23). With our first review of the literature in 2015 (22), we found that whilst the literature was conflicting (32 of 58 measures showed NLMF), there was evidence that NLMF effects were more apparent when the lower body (e.g., quadriceps) was tested, isometric and

DOI: 10.4324/9781032709086-7

cyclical fatigue interventions had a greater incidence and magnitude of NLMF, and fatiguing or extended duration testing protocols exhibited greater NLMF. Hence, we pursued these issues in subsequent research. Six years later, we published a meta-analysis of 52 studies (278 measures), which did not support the existence of a "general" NLMF effect (e.g., strength and power deficits); but there was evidence for endurance-based (i.e., time to task failure) NLMF impairments (6). However, another interesting gap arose from the "shoulders of giants" and we wondered whether there was crossover, non-local, or global effects of stretching. Could you stretch one muscle and have increased range of motion (ROM) or flexibility in other joints and muscles?

Global Stretching Effects on Range of Motion

Lo and behold, there were acute global (non-local and crossover) effects of unilateral stretching. Some of the first studies demonstrated that if one hamstrings was stretched then the contralateral hamstrings (hip flexion ROM) experienced greater flexibility (16), if participants stretched their hip abductors (groin), they experienced greater shoulder (horizontal abduction) ROM and if they stretched their shoulders, then there was an improvement in hip flexion (hamstrings) ROM (8).

We also wondered about the possibility of cross education, stretch training effects. Cross education has been demonstrated since 1894 by Scripture (46) with only two participants (Miss Smith and Miss Brown) who showed improvements in contralateral muscular control and power after unilateral practice. Since then, there have been innumerable studies that show a cross transfer of training effects for motor skills, muscle strength, power, and other attributes in all segments of the population (18, 19, 24, 31). We compared single versus twice daily unilateral stretch training sessions on hip flexion ROM of the stretched and contralateral, non-stretched legs. Both one and two per day stretch training sessions over two weeks improved both active static (5.9 percent vs 5.2 percent ROM improvements for stretched and contralateral respectively) and ballistic (5.01 percent vs. 2.3 percent ROM improvements for stretched and contralateral respectively) ROM (12). Nakamura et al. (40) compared the cross-education effects of unilateral stretch training with four weeks of high- vs. low-intensity stretching on dorsiflexion ROM, muscle stiffness, and architecture. They found significant, moderate magnitude, contralateral dorsiflexion ROM increases with the high intensity stretch group but not the low intensity group. There were no changes in muscle stiffness or architecture.

You would not expect everybody to agree, would you? In contrast, a ten-week stretch training programme of the plantar flexors (three days/week, four repetitions of 30 seconds each with 30 seconds rest between stretches) did not induce cross transfer improvements in ROM, but did produce significant strength gains (41). Similarly, the Konrad group did not find global (non-local effects on heterologous muscles or regions) improvements in shoulder extension ROM after seven weeks of static stretching and foam rolling on the sole of the foot (28) or with seven weeks of pectoralis muscle static stretching (both studies intervened three days/week with three exercises of five minutes each) on ankle dorsiflexion ROM (29).

You will notice in this second edition, that many meta-analyses are reported. During the COVID-19 pandemic, we and others could not collect data as all labs were closed due to the pandemic. In order to stay busy, stay relevant, and keep contributing to science, we published a number of meta-analyses. Meta-analyses are a great way to survey the literature and quantitatively evaluate what has worked and not worked and then highlight new research directions to pursue. One of these meta-analyses (4) showed that an acute bout of unilateral passive static stretching induced moderate magnitude improvements in passive ROM with non-local, non-stretched joints. Participant sex, trained state, stretching intervention intensity, and duration did not moderate these results. We also noticed that more than 240 seconds of stretching induced large magnitude increases in non-local ROM compared with moderate magnitude enhancements with shorter (< 120 seconds) durations of stretching. It was emphasized earlier in this text that static stretching should not exceed 60 seconds per muscle group as part of a warm-up, otherwise there could be subsequent performance impairments.

Global Stretching Effects on Performance

But, as the early acute stretching research from our and other labs highlighted that prolonged static stretching without a full warm-up (addition of dynamic activities before and after static stretching) could induce performance impairments (7, 9, 15, 26), we wondered whether this was possible in a non-stretched, non-exercised muscle. As some research demonstrated declines in electromyography (EMG), EMG/M-wave (muscle action potential wave) ratios and voluntary activation (e.g., interpolated twitch technique, volitional wave), it was suggested that there may be a reduction in central (neural efferent) drive (49, 50). Furthermore, muscle stretching can diminish persistent inward currents (PIC) at the motoneurons due to reduced muscle spindle facilitation (51). PICs can increase the gain of the motoneuron output by amplifying the synaptic inputs to the motor unit facilitating force output (51). Would these deficits occur solely with the motoneurons of the stretched muscle or would they spread to contralateral motoneurons?

In addition, it has been postulated that long duration fatigue interventions and perhaps prolonged stretching may have physio-psychological implications, which could result in a mental energy deficit when experiencing stretching to the point of discomfort or pain. This mental fatigue, could reduce the focus, attention, and concentration, needed to achieve maximal or near maximal neuromuscular activation (22), especially with prolonged or repeated contractions. Steele (47) discusses the "perception of effort" which is defined as the individual's perception of the required effort to accomplish a specific task or set of tasks,. It is possible that the prolonged stretching to the point of discomfort may have elevated the perception of effort involved in the subsequent task, which could contribute to global or non-local performance impairments.

A systematic review of the limited literature investigating the global effects of prolonged static stretching (mean stretching characteristics: 6.3 ± 2 repetitions of 36.3 ± 7.4 seconds with 19.3 ± 5.7 seconds of recovery between stretches) revealed that both the stretched and contralateral, non-stretched, muscles revealed small magnitude force deficits (5). However, the power of these findings was not strong as the frequency of studies with these effects were similar with three measures demonstrating deficits, and four measures showing trivial changes.

Mechanisms Underlying Global Stretching Effects

Non-local or global ROM improvements could be attributed to increased stretch or pain tolerance (32, 33, 35, 48). Increased stretch tolerance proposes that the musculotendinous unit can tolerate greater stress without a change in tension for a given length (33, 34). Augmented stretch tolerance is often promoted as an underlying mechanism for the increase in the ROM of the stretched muscle (7, 10, 15). We unilaterally treated the hamstrings with a menthol based topical analgesic to decrease pain sensitivity and discovered an increased passive static and ballistic hip flexion ROM in both the treated and contralateral limbs (52). Whether the increased pain (stretch) tolerance is primarily psychological or physiological (e.g., diffuse noxious inhibitory control theory suggests a release of endorphins and enkephalins in response to pain), the effects would be felt throughout the body (global) and not just directed to the affected (stretched) region, limb, or joint.

Other possibilities may involve a downregulation of sympathetic excitatory nerve stimulation (4). The goal of pain perception is to alert the physiological system to potential threats or damage and therefore upregulate the sympathetic nervous system (fight or flight) (38). However, with prolonged stretching near or at the point of discomfort, the perception that stretching would be an injury threat would diminish and therefore the fight or flight response (increased sympathetic excitation) would be downregulated. This would result in muscle relaxation permitting greater extensibility. Once again, sympathetic stimulation or parasympathetic relaxation is a global response and all muscles would experience this effect and not just the stretched muscles.

Further muscle relaxation may be achieved with sustained static stretching (e.g., 30 seconds or more) by attenuating motoneuron spindle reflexive (excitatory) activity (3, 20). As early as 1959 (42), stretch-induced changes in the myotatic reflex and inverse myotatic reflex were associated with reciprocal modifications in contralateral motoneurons of cat hind limb muscles. Type I and II muscle spindle afferents not only innervate spinal motoneurons (44) but also connect to the somatosensory and the primary motor cortex (43). As the signals project to the cortex, the decrease in reflex derived excitability would have a global, full body reach. The decreased sympathetic nervous system stimulation and disfacilitation of excitatory spindle afferents (neural inhibition) could potentially increase non-local ROM while also potentially impairing performance. However, a study by Hadjizadeh Anvar et al. (2) reported contralateral ROM improvements with no significant spinal (H-reflex) and corticospinal excitability (motor evoked potentials with transcranial magnetic stimulation) changes suggesting that afferent excitability of the spinal motoneurons and corticospinal excitability may not play a role in global muscle's ROM increases. Contradicting the Hadjizadeh Anvar findings, Masugi et al. (37) reported that plantar flexors' muscle stretching (dorsiflexion stretch) demonstrated inhibition of monosynaptic spinal reflexes, for both the stretched and non-stretched muscles of the ipsilateral leg. More work is needed in this area to determine possible neural effects.

It has also been suggested that myofascial chains may play a role (30, 53, 54). The concept of myofascial chains posits that there is a continuous transmission of tension with the myofascia, which covers and extends from lower to upper body muscles (30). This continuity can either contribute to force transmission or if the myofascia becomes more extensible or compliant in one area, it can release tension in other areas of the body. For example, lower limb stretching provided increases in cervical ROM and was attributed to myofascial chains (13, 54). Wilke and colleagues (53) suggested that tissue strain and stress can be transferred from one body region to another, accentuating global connectivity. However, Wilke et al. (54) did not directly measure changes in fascial structural and thus stretch tolerance or reduced sympathetic excitation might have been major contributors. As the force transfer with myofascial chains or meridians predominately align longitudinally (11, 17), obliquely (39) as well as transversal to synergists and antagonists (25), these myofascial chains may not contribute substantially to an increased extensibility of a contralateral homologous muscle (i.e., quadriceps to quadriceps). Furthermore, the aforementioned two Konrad studies (28, 29) did not reveal a non-local increase in ROM when the stretching occurred either at the shoulders (test ankle dorsiflexion ROM) or the sole of the foot (test shoulder extension ROM). So, although we have seen improvements when stretching the hamstrings positively affecting the shoulders and vice versa, if there is an effect of myofascial chains it may not be able to significantly transfer effects over this greater cephalocaudal distance.

Global Effects of Foam Rolling

Similar to the crossover or global effects of stretching, foam rolling has also demonstrated increased contralateral ROM following an acute (single session) bout of ipsilateral (unilateral) foam rolling. The evidence is not as comprehensive with chronic foam rolling training effects (27). Once again, similar to global effects of stretching, reduction in pain perception is suggested as the most probable mechanism for contralateral ROM increases (27). As further evidence of global decreases in pain perception, unilateral rolling of the plantar flexors provided similar increases in pain pressure threshold (less pain sensitivity) in one study that monitored calf muscle tender points (1) and another study that examined the degree of pain with high frequency evoked tetanic contractions (14). In both cases, without even touching the affected muscle, all we had to do was roll the contralateral muscle and pain was reduced. With lower pain sensitivity you can push yourself through a greater ROM with less discomfort.

In terms of global or contralateral performance effects, the Konrad et al. scoping review (27) reported no changes in muscle performance (i.e., maximum voluntary isometric contraction, maximum voluntary dynamic contraction) after an acute unilateral foam rolling bout (just 4 studies with seven measures). However, one of the four studies found substantial (large magnitude) deficits in rate of force development of the contralateral limb following unilateral foam rolling of the ipsilateral leg (55). Ye et al. (55) postulated that the impairment might be related to a decrease in motor unit recruitment. The only two studies evaluating chronic unilateral foam rolling training found moderate magnitude enhancements. With the small number of studies in this scoping review Professor Konrad warned that their findings should be interpreted with caution.

Summary

Stretching a particular muscle or joint can have global ramifications for the body with increases in flexibility or ROM not only with the stretched muscle and joint but other non-stretched muscles. This may be attributed to a global increase in stretch (pain) tolerance, attenuated sympathetic stimulation or perhaps decreased reflex excitation. Non-local or global stretching effects would be important for training and rehabilitation, when a musculotendinous injury prevents training or activity. However, the finding of greatest non-local passive ROM with more than four minutes of stretching should not be prescribed for warm-ups but for distinct stretch training sessions. There is evidence that performance impairments associated with prolonged static stretching can also be transferred to contralateral muscles. Thus, unilateral stretching could maintain or induce greater contralateral flexibility without exacerbating the injury to the affected muscles or tendons.

References

1. Aboodarda, S.J., Spence, A.J., and Button, D.C.Pain pressure threshold of a muscle tender spot increases following local and non-local rolling massage. *BMC Musculoskelet Disord* 16: 265, 2015.
2. Anvar, S.H., Granacher, U., Konrad, A., Alizadeh, S., Culleton, R., Edwards, C., Goudini, R., and Behm, D.G.Corticospinal excitability and reflex modulation in a

contralateral non-stretched muscle following unilateral stretching. *Eur J Appl Physiol*, 2023.

3. Avela, J., Kyröläinen, H., and Komi, P.V.Altered reflex sensitivity after repeated and prolonged passive muscle stretching. *Journal of Applied Physiology* 86: 1283–1291, 1999.
4. Behm, D.G., Alizadeh, S., Anvar, S.H., Drury, B., Granacher, U., and Moran, J. Non-local Acute Passive Stretching Effects on Range of Motion in Healthy Adults: A Systematic Review with Meta-analysis. *Sports Med* 51: 945–959, 2021.
5. Behm, D.G., Alizadeh, S., Drury, B., Granacher, U., and Moran, J.Non-local acute stretching effects on strength performance in healthy young adults. *Eur J Appl Physiol* 121: 1517–1529, 2021.
6. Behm, D.G., Alizadeh, S., Hadjizedah Anvar, S., Hanlon, C., Ramsay, E., Mahmoud, M.M.I., Whitten, J., Fisher, J.P., Prieske, O., Chaabene, H., Granacher, U., and Steele, J.Non-local Muscle Fatigue Effects on Muscle Strength, Power, and Endurance in Healthy Individuals: A Systematic Review with Meta-analysis. *Sports Med* 51: 1893–1907, 2021.
7. Behm, D.G., Blazevich, A.J., Kay, A.D., and McHugh, M.Acute effects of muscle stretching on physical performance, range of motion, and injury incidence in healthy active individuals: a systematic review. *Appl Physiol Nutr Metab* 41: 1–11, 2016.
8. Behm, D.G., Cavanaugh, T., Quigley, P., Reid, J.C., Nardi, P.S., and Marchetti, P.H. Acute bouts of upper and lower body static and dynamic stretching increase non-local joint range of motion. *Eur J Appl Physiol* 116: 241–249, 2016.
9. Behm, D.G. and Chaouachi, A.A review of the acute effects of static and dynamic stretching on performance. *Eur J Appl Physiol* 111: 2633–2651, 2011.
10. Behm, D.G., Kay, A.D., Trajano, G.S., and Blazevich, A.J.Mechanisms underlying performance impairments following prolonged static stretching without a comprehensive warm-up. *Eur J Appl Physiol* 121: 67–94, 2021.
11. Benetazzo, L., Bizzego, A., De Caro, R., Frigo, G., Guidolin, D., and Stecco, C.3D reconstruction of the crural and thoracolumbar fasciae. *Surg Radiol Anat* 33: 855–862, 2011.
12. Caldwell, S.L., Bilodeau, R.L.S., Cox, M.J., and Behm, D.G.Twice daily, self-administered band stretch training improves quadriceps isometric force and drop jump characteristics. *Journal of Sport Science and Medicine*, 2019.
13. Calgaro, J., Bonaldi, L., Sposta, S.M., Fede, C., Stecco, A., Pirri, C., and Stecco, C.Effects of Lower Limbs Stretching on the Neck Range of Motion: Preliminary Evidence for Myofascial Sequence? *International Journal of Orthopedics and Rehabilitation* 9: 8–14, 2023.
14. Cavanaugh, M.T., Döweling, A, Young, J.D., Quigley, P.J., Hodgson, D.D., Whitten, J.H., Reid, J.C., Aboodarda, S.J., and Behm, D.G.An acute session of roller massage prolongs voluntary torque development and diminishes evoked pain. *Eur J Appl Physiol* 117: 109–117, 2017.
15. Chaabene, H., Behm, D.G., Negra, Y., and Granacher, U.Acute Effects of Static Stretching on Muscle Strength and Power: An Attempt to Clarify Previous Caveats. *Front Physiol* 10: 1468, 2019.
16. Chaouachi, A., Padulo, J., Kasmi, S., Othmen, A.B., Chatra, M., and Behm, D.G. Unilateral static and dynamic hamstrings stretching increases contralateral hip flexion range of motion. *Clin Physiol Funct Imaging* 37: 23–29, 2017.

17. Eng, C.M., Pancheri, F.Q., Lieberman, D.E., Biewener, A.A., and Dorfmann, L. Directional differences in the biaxial material properties of fascia lata and the implications for fascia function. *Ann Biomed Eng* 42: 1224–1237, 2014.
18. Green, L.A. and Gabriel, D.A.The cross education of strength and skill following unilateral strength training in the upper and lower limbs. *J Neurophysiol* 120: 468–479, 2018.
19. Green, L.A. and Gabriel, D.A.The effect of unilateral training on contralateral limb strength in young, older, and patient populations: a meta-analysis of cross education. *Physical Therapy Reviews*, 2018.
20. Guissard, N., Duchateau, J., and Hainaut, K.Mechanisms of decreased motoneurone excitation during passive muscle stretching. *Experimental Brain Research* 137: 163–169, 2001.
21. Halperin, I., Aboodarda, S.J., and Behm, D.G.Knee extension fatigue attenuates repeated force production of the elbow flexors. *Eur J Sport Sci* 14: 823–829, 2014.
22. Halperin, I., Chapman, D.W., and Behm, D.G.Non-local muscle fatigue: effects and possible mechanisms. *Eur J Appl Physiol* 115: 2031–2048, 2015.
23. Halperin, I., Copithorne, D., and Behm, D.G.Unilateral isometric muscle fatigue decreases force production and activation of contralateral knee extensors but not elbow flexors. *Appl Physiol Nutr Metab* 39: 1338–1344, 2014.
24. Hellebrandt, F.A.Cross education: Ipsilateral and contralateral effects of unimanual training. *JAP* 4: 136–144, 1951.
25. Huijing, P.A., van de Langenberg, R.W., Meesters, J.J., and Baan, G.C.Extramuscular myofascial force transmission also occurs between synergistic muscles and antagonistic muscles. *J Electromyogr Kinesiol* 17: 680–689, 2007.
26. Kay, A.D. and Blazevich, A.J.Effect of acute static stretch on maximal muscle performance: a systematic review. *Med Sci Sports Exerc* 44: 154–164, 2012.
27. Konrad, A., Nakamura, M., Warneke, K., Donti, O., and Gabriel, A.The contralateral effects of foam rolling on range of motion and muscle performance. *Eur J Appl Physiol* 123: 1167–1178, 2023.
28. Konrad, A., Reiner, M., Manieu, J., Fischer, J., Schopflin, A., Tilp, M., and Behm, D. G.The non-local effects of 7-week foot sole static stretching and foam rolling training on shoulder extension range of motion. *Front Sports Act Living* 5: 1335872, 2023.
29. Konrad, A., Reiner, M., Manieu, J., Fischer, J., Schopflin, A., Tilp, M., and Behm, D.G.Seven weeks of pectoralis muscle stretching does not induce non-local effects in dorsiflexion ankle range of motion. *European Journal of Sports Science*: 1–7, 2024.
30. Krause, F., Wilke, J., Vogt, L., and Banzer, W.Intermuscular force transmission along myofascial chains: a systematic review. *J Anat* 228: 910–918, 2016.
31. Lee, M. and Carroll, T.J.Cross education: possible mechanisms for the contralateral effects of unilateral resistance training. *Sports Med* 37: 1–14, 2007.
32. Magnusson, S.P. and Renstrom, P.The European College of Sports Sciences Position statement: The role of stretching exercises in sports. *European Journal of Sport Science* 6: 87–91, 2006.
33. Magnusson, S.P., Simonsen, E.B., Aagaard, P., Boesen, J., Johannsen, F., and Kjaer, M.Determinants of musculoskeletal flexibility: viscoelastic properties, cross-sectional area, EMG and stretch tolerance. *Scand J Med Sci Sports* 7: 195–202, 1997.
34. Magnusson, S.P., Simonsen, E.B., Aagaard, P., Gleim, G.W., McHugh, M.P., and Kjaer, M.Viscoelastic response to repeated static stretching in the human hamstring muscle. *ScandJMedSciSports* 5: 342–347, 1995.

35. Magnusson, S.P., Simonsen, E.B., Aagaard, P., Sorensen, H., and Kjaer, M.A mechanism for altered flexibility in human skeletal muscle. *Journal of Physiology* 497 (Pt 1): 291–298, 1996.
36. Martin, P.G. and Rattey, J.Central fatigue explains sex differences in muscle fatigue and contralateral cross-over effects of maximal contractions. *Pflügers Archiv: European journal of physiology* 454: 957–969, 2007.
37. Masugi, Y., Obata, H., Inoue, D., Kawashima, N., and Nakazawa, K.Neural effects of muscle stretching on the spinal reflexes in multiple lower-limb muscles. *PLoS One* 12: e0180275, 2017.
38. Moayedi, M. and Davis, K.D.Theories of pain: from specificity to gate control. *J Neurophysiol* 109: 5–12, 2013.
39. Myers, T.W. *Anatomy trains: Myofascial meridians for manual and movement therapists.* Edinburgh: Churchill Livingstone Publishers, 2001.
40. Nakamura, M., Yoshida, R., Sato, S., Yahata, K., Murakami, Y., Kasahara, K., Fukaya, T., Takeuchi, K., Nunes, J.P., and Konrad, A.Cross-education effect of 4-week high- or low-intensity static stretching intervention programs on passive properties of plantar flexors. *J Biomech* 133: 110958, 2022.
41. Nelson, A.G., Kokkonen, J., Winchester, J.B., Kalani, W., Peterson, K., Kenly, M. S., and Arnall, D.A.A 10-week stretching program increases strength in the contralateral muscle. *J Strength Cond Res* 26: 832–836, 2012.
42. Perl, E.R.Effects of muscle stretch on excitability of contralateral motoneurones. *J Physiol* 145: 193–203, 1959.
43. Phillips, C.G., Powell, T.P., and Wiesendanger, M.Projection from low-threshold muscle afferents of hand and forearm to area 3a of baboon's cortex. *J Physiol* 217: 419–446, 1971.
44. Prochazka, A. and Ellaway, P.Sensory systems in the control of movement. *Compr Physiol* 2: 2615–2627, 2012.
45. Rattey, J., Martin, P.G., Kay, D., Cannon, J., and Marino, F.E.Contralateral muscle fatigue in human quadriceps muscle: evidence for a centrally mediated fatigue response and cross-over effect. *Pflügers Archiv: European Journal of Physiology* 452: 199–207, 2006.
46. Scripture, E.W., Smith, T.L., and Brown, E.M.On the education of muscular control and power. *Studies of the Yale Psychology Laboratory* 2: 114–119, 1894.
47. Steele, J.What is (perception of) effort? Objective and subjective effort during task performance. *PsyArχiv*, 2020.
48. Stove, M.P., Hirata, R.P., and Palsson, T.S.Muscle stretching − the potential role of endogenous pain inhibitory modulation on stretch tolerance. *Scand J Pain* 19: 415–422, 2019.
49. Trajano, G.S., Seitz, L., Nosaka, K., and Blazevich, A.J.Contribution of central vs. peripheral factors to the force loss induced by passive stretch of the human plantar flexors. *J Appl Physiol (1985)* 115: 212–218, 2013.
50. Trajano, G.S., Seitz, L.B., Nosaka, K., and Blazevich, A.J.Can passive stretch inhibit motoneuron facilitation in the human plantar flexors? *J Appl Physiol (1985)* 117: 1486–1492, 2014.
51. Trajano, G.S., Taylor, J.L., Orssatto, L.B.R., McNulty, C.R., and Blazevich, A.J. Passive muscle stretching reduces estimates of persistent inward current strength in soleus motor units. *J Exp Biol* 223, 2020.

52. Whalen, A., Farrell, K., Roberts, S., Smith, H., and Behm, D.G.Topical Analgesic Improved or Maintained Ballistic Hip Flexion Range of Motion with Treated and Untreated Legs. *J Sports Sci Med* 18: 552–558, 2019.
53. Wilke, J., Krause, F., Vogt, L., and Banzer, W.What Is Evidence-Based About Myofascial Chains: A Systematic Review. *Arch Phys Med Rehabil* 97: 454–461, 2016.
54. Wilke, J., Vogt, L., Niederer, D., and Banzer, W.Is remote stretching based on myofascial chains as effective as local exercise? A randomised-controlled trial. *J Sports Sci* 35: 2021–2027, 2017.
55. Ye, X., Killen, B.S., Zelizney, K.L., Miller, W.M., and Jeon, S.Unilateral hamstring foam rolling does not impair strength but the rate of force development of the contralateral muscle. *Peer J* 7: e7028, 2019.

7 Recommendations for Stretching Prescription

If the goal of the stretching session is to induce a plastic increase in range of motion (ROM), then the stretching should not be performed as part of a warm-up or a cooldown. Stretching during these periods has different objectives and the possibility of negative performance implications for performance if the stretching is prolonged and performed in isolation before an activity (see Chapter 8). Details about stretching as part of the warm-up will be provided in a following section.

Stretching Duration

Almost any duration of stretching can improve ROM. In almost every study, you read, the very act of measuring a joint ROM will improve elastic flexibility. In one of our studies, we found we needed to pre-test individuals at least three times to ensure the ROM did not increase as a result only of the testing (33). Thus, a single stretch of less or equal to five seconds may improve ROM (65). That is why every stretching study needs a control condition or group! Whereas Roberts and Wilson (65) showed that nine stretches of five seconds provided similar increases in passive ROM as three stretches of 15 seconds, the 15 second stretches had a significantly better effect on active ROM than the 5 second duration stretches. Earlier studies by Bandy and Irion (4) suggested that 30–60 seconds of static stretching was more effective than 15 seconds to increase passive ROM. A static stretch training study had participants train three times a week for five weeks and found significant improvements in ROM with five seconds of stretch training but 15 seconds of stretching provided greater improvements (65). Other researchers also support using greater than 30 seconds of static stretching to achieve the greatest ROM (17, 27). In a later study by Bandy et al. (5) they indicated that one 30 second static stretch per day increased hamstrings ROM, but there were no differences with increased repetitions or frequency. In an animal study using rabbits (75), it was found that the greatest length changes in the MTU occurred within the first four stretches. In humans, two 30-second static stretches were needed to significantly decrease musculotendinous stiffness of the plantar flexors (68). There were no further decreases with 3–4 stretches of

DOI: 10.4324/9781032709086-8

the same duration. While we are obviously not the same as rabbits, the findings with these warm-blooded mammalian cousins in association with some of the human studies tend to provide the same message: there is no need for an exaggerated duration of stretching to obtain acute optimal increases in ROM. Thomas et al. (76) in their meta-analysis recommended a minimum duration of five minutes per week for each muscle group. Once again referring back to the Konrad et al.'s meta-analysis (42), they reported no stretch training ROM differences based on stretching volume or intensity.

Stretching Intensity

Many studies utilize static stretches that elongate the musculotendinous unit or joint to the point of discomfort (pain) or near the point of discomfort (6, 7, 8, 9, 10). Mechanically, this stretch point would be referred to as the elastic limit. The elastic limit is the minimum amount of stress placed on a tissue to elicit permanent strain. Exceeding the elastic limit will result in the tissue not returning to its original length after the stretch (1). At this point, musculotendinous strains or ligamentous sprain injuries could occur. Alter (1) expands on the strength training overload principle (28, 29, 30), transferring the concept to flexibility training as the overstretching principle. The overstretching principle is described as "when the body is regularly stimulated by an increasingly intense stretching programme beyond the homeostatic level, it will respond with an increased ability to stretch" (p. 145). The question arises, is it necessary to reach the elastic limit or point of discomfort to achieve plastic or semi-permanent changes in flexibility?

A number of acute studies have shown that submaximal intensity stretches provide similar ROM benefits as near maximal point of discomfort stretches (40, 41, 53, 85). We had subjects stretch at 100 percent, 75 percent, and 50 percent of the point of discomfort but did not find any significant difference in the flexibility test (stoop and reach). All conditions improved by approximately 12 percent (9). In contrast, another study reported that static stretching at 85–100 percent of maximum stretch intensity provided greater ROM than stretching at 60 percent (80). Other studies have compared high force-short duration to low force-long duration stretching and report that high force stretches emphasize elastic tissue deformation that shortly returns to its original length, whereas low force, prolonged stretching enhances plastic or semi-permanent changes in tissue length (48, 73, 81, 82). Apostopoulos (2) recommended stretching at 30–40 percent of perceived exertion. As a reminder from the last chapter, the Konrad et al. meta-analysis (42) revealed no ROM differences based on chronic stretch training volume, intensity, or weekly frequency. Stretching to the point of pain could be counterproductive. A typical response to pain or discomfort or distress is to adopt a stiffening strategy (16): that is to contract both the agonist and antagonist muscles in order to protect them from possible physical insults. Hence, while the individual is trying to lengthen the muscle, the central nervous system is trying to shorten the muscle. So, there is no need for masochism when stretching, pain is not a necessity. Furthermore, stretching to the point of discomfort or elastic limit for a recently injured or fatigued tissue could strain (muscles and tendons) or sprain (ligaments) the tissue. Thus, during rehabilitation or after an intense, fatiguing, training session or match, high intensity stretching should not be pursued. Another point to consider is that pain is highly subjective to the individual. Hence, telling a person with a high pain threshold to stretch to the point of discomfort will place much greater stress on the tissues than for someone who has a lower tolerance to pain. It is much safer and reportedly equally effective to stretch below pain tolerance!

Continuous stretching (no rest periods) permits higher stretching intensities (31). However as mentioned, higher stretching intensities may not provide any extra ROM benefits. In fact, one study reported that intermittent stretching reduced peak torque (i.e., strength) more than continuous stretching, which were strongly associated with a depression in central drive (77).

Optimal Time of Day to Stretch

When is the best time of day to stretch? There are diurnal (time of day) variations in performance. Typically, most people perform maximal strength and power activities better in the afternoon compared to the morning. Dynamic stretching has been shown to counteract the lower vertical jump heights often found in the morning (18). Dr. Stuart McGill, a highly respected biomechanist (with one of the best moustaches around) from the University of Waterloo suggests that extensive stretching in the morning is counterproductive. Vertebral discs are infused with fluid known as the nucleus pulposa. After a night of being horizontal during sleep, the disc's nucleus pulposa become more hydrated. The gravity-induced fluid loss owing to a day of upright posture is replaced. This modulation in fluid alters the stresses on the disc throughout the day. Specifically, stress is highest following bedrest then diminishes over the subsequent few hours. Stretching the lower back in the morning with the discs expanded can increase the chances for disc herniation. Passive tissues of the back can be injured by bending over to pick up a pencil especially if the individual is unstable. Thus, attempting to touch the toes whether in the morning or any other time of the day can lead to injury. In fact, McGill proposes that a more flexible back actually increases the risk of future back problems. He states there is trade-off between mobility and stability and that balance is specific to the individual based on prior injuries, age, training objectives and other factors (55).

Stretching Frequency (Days/week)

Should we stretch every day or alternate days? The Konrad et al. meta-analysis (42) did not find any significant differences in ROM associated with the weekly frequency. Perhaps examining and contrasting with the strength training literature would be helpful on this issue. Resistance training the same muscle on subsequent days is not advised since the muscle needs time to recover. If the intensity is sufficient, overload resistance stress can degrade muscle protein and activate a number of receptors and pathways for increased protein synthesis. This overcompensation to the stress can result in increased muscle strength and hypertrophy (43, 44, 45). However, without a sufficient recovery period (typically 48–72 hours), the muscle pathways continue to emphasize protein degradation rather than the synthesis of protein resulting in decreased strength and atrophy. For the most extreme example, contemplate the muscle hypertrophy of a bodybuilder to the muscle atrophy of a war-time concentration camp prisoner. Both individuals receive an overload stress on the muscle, but the bodybuilder builds appropriate rest periods (and of course proper nutrition) into the programme whereas the prisoner works hard every day without the chance for overcompensation. The difference between resistance training and stretching is the lower overload intensity and lack of protein degradation with stretching. Most stretching programmes would not lead to similar protein degradation or depletion of muscle glycogen stores as resistance training and thus there would not be a need for a prolonged recovery period. There is ample evidence of individuals who stretch every day providing significant improvements in flexibility without any negative consequences. It is possible that extreme flexibility programmes as seen with some gymnasts, dancers and other similar athletes may damage muscle and connective tissue and might be more effective with alternate day stretching. However, this type of comparative research has not been conducted. Flexibility training studies have successfully improved ROM with two (46), three (23), four (86), five (12), and seven (47, 52, 61, 67, 84) days/week flexibility training programmes. As can be seen from the number of citations, daily stretching has been commonly used in the literature. Thus, for the vast majority of people who stretch, daily stretching should provide substantial improvements in flexibility. As mentioned previously, stretching only 1 day / week can sustain previously attained flexibility training gains (79). Since no extensive direct comparison studies (i.e., one study that directly compared one vs. three vs. five vs. seven days of stretching per week) have been conducted, we cannot state whether daily stretching actually provides better flexibility improvements than five days/week, three days/week or other frequencies. However, the Thomas et al. (76) meta-analysis comparing the results of studies using differing stretching frequencies suggested that stretching at least five days per week (with a minimum five minutes per muscle group per week) provided the most beneficial increases in ROM. Furthermore, while Alter (1) suggests that stretching once a day would maintain flexibility and "empirical evidence suggests that stretching at least twice a day is preferable" (p. 154), there is scant evidence regarding whether stretching multiple times per day substantially enhances the improvement in flexibility.

Pre- versus post-workout stretching

Should we stretch prior to or after a workout? There is an entire section later in the text on the effects of pre-activity stretching during a warm-up, so we will leave that discussion till then. A common practice is to perform stretching exercises after a workout with the rationale being that the muscles and connective tissue are warm and the viscosity is low. This decreased viscosity is certainly an advantage for achieving greater muscle and tendon lengths, but dependent upon the prior activity, the muscles may be fatigued as well. Hence, attempting to stretch musculotendinous tissue to the maximum point of discomfort in order to achieve greater extensibility could lead to tissue strains if the tensile strength of the tissue is compromised by fatigue. Therefore, post-exercise stretching especially if fatigue is induced should involve low to moderate intensity stretching, so as not to overload the tension on the muscles and tendons. This stretching period should be more in the relaxation mode than to increase musculotendinous extensibility!

Stretching for Relaxation

Static stretching can physiologically relax an individual. Static stretching (five repetitions x 1 minute each of triceps surae) changed the predominance of the autonomic nervous system to a greater parasympathetic neural influence during the stretch and continued for four minutes following the stretch (38). This effect can last substantially longer as another study found the greater parasympathetic influence returning to pre-stretch levels 30 minutes after stretching (three stretches x 30 seconds each) (25). Relaxation can be an important aspect of exercise recovery. High stress whether it be physiological or psychological can increases cortisol (78). Cortisol, a catabolic hormone, increases protein degradation, increases the metabolism of protein, fat, and carbohydrates and suppresses the immune system (32). With recovery, we basically want the opposite actions to occur. Thus, while high intensity stretching is not recommended after a workout with the objective of increasing ROM, stretching in order to relax would be a strong recommendation after taxing physical activity. In addition to stretching, a focus on breathing patterns can affect parasympathetic activation. Yoga, of course has developed and focused on a variety of ventilatory strategies for centuries.

Ventilatory Effects on Stretching

A growing body of evidence supports the belief that yoga benefits physical and mental health via down-regulation of the hypothalamic-pituitary-adrenal axis and the sympathetic nervous system (3). With yoga, the emphasis is often on deep, rhythmic, consistent, diaphragmatic and nasal breathing to relax the individual (21). In a more relaxed state, it is believed that the individual will be able to achieve greater ROM. Unpublished data from our lab showed that unilateral nasal breathing could affect heart rate (Kliger et al.) While walking on a treadmill for 10 minutes, with unilateral left nostril breathing, there were significant increases in heart rate, systolic and diastolic blood pressures. Post-treadmill walking, flexibility was tested. A significant decrease in flexibility was found. In contrast, greater flexibility was achieved, with decreased heart rate and blood pressure during walking after unilateral right nasal breathing. These results correspond with Yoga Tradition. Energy flow through "ida" (during left nasal breathing practice) is supposed to be 'heat dissipating (cooling)' whereas energy flow through "pingala" (during right nasal breathing practice) is 'heat generating'. Hence, there may be a nostril laterality affecting the autonomous nervous system differentially (22), with left nostril breathing emphasizing the parasympathetic influence.

Similar to breathing or ventilatory patterns when resistance training, the breathing pattern can have an effect on stretching especially when performing a trunk forward flexion action. Expiring during trunk forward flexion is typically recommended as a full inspiration tends to contract the erector spinae muscles (15, 64) and expand the thoracic cage (ribs), which would detract from attempting to fully flex (1). In contrast, Hamilton and colleagues (34) found greater ROM and lower electromyography (EMG) activity of the rectus abdominus, external obliques, lower abdominal stabilizers, and lower erector spinae of women with breathing techniques that emphasized larger inhalations. There was no effect of different breathing techniques on the male participants. The greater joint stiffness of men (13) may have contributed to lack of ventilatory effects. McHugh et al. (56) reported that mechanical factors contribute about 80 percent to the trunk forward flexion ROM, and thus the intrinsic male mechanical stiffness may have overcome neural or mechanical ventilatory effects. For both sexes in the study, EMG activity while inhaling before the stretch was not significantly lower, which tends to indicate that relaxed trunk muscle activity is not the single most important factor for attaining greater flexibility, contradicting some earlier studies (15, 64). Furthermore, pulmonary stretch reflexes inhibit sympathetic nerve activity at higher lung tidal volumes (71), thus there could have been lower sympathetic nerve activity for the pre-stretch inhale, inhale-during stretch, and hypoventilation conditions in the Hamilton study. Hamilton recommended that women should inhale at a slow frequency (hypoventilation) before and maintain that inhalation during the stretch. In contrast to the earlier studies cited above, they found that forceful exhalations actually increased EMG activity, thus inhibiting hip flexion ROM.

Combining Stretching with Muscular Contractions or Massage

Other techniques might be applied during a stretch to augment ROM increases. Implementing muscular contractions during a static stretch (active static stretch) might provide some greater benefit early in a training programme (i.e., four weeks) but no additional benefit over a longer time period compared to passive stretching (i.e., eight weeks) (26). Another technique that can augment the stretching effect is massage of the tendon. In a study from our laboratory (37), friction massage was applied to the hamstrings tendon for either ten or 30 seconds prior to testing for hip flexion (hamstrings) ROM. Both durations provided 6–7 percent increases in ROM. The heat-induced friction of the massage would have decreased visco-elasticity and perhaps activated cutaneous and myofascial afferents helping to inhibit reflex-induced contractions. Massaging the tendon either before or during the stretch can augment the effectiveness.

Effect of Temperature on Stretching

Another consideration would be to ensure the tissues are at a higher temperature (hyperthermia) when stretching. Increased tissue temperatures decrease tissue stiffness and increase extensibility (48, 59, 62). Tendon temperatures over 39.4°C (103F) can augment plastic elongation (48, 50), while tissue temperatures over 40°C (104F) can enhance the viscous stress relaxation for collagen protein also leading to greater plastic deformation (54, 62, 63). There are conflicting studies regarding hyperthermic applications with increased ROM found following hot baths (72), hot packs (10 minutes) (51), ultrasound (83), and diathermy (60). But a lack of ROM augmentation has been reported after applying moist heat (70), or an electric heating pad (20 minutes at 43°C) (36).

There is controversy over whether a hypothermic (cold) stretch provides the best tissue lengthening (35, 51), or whether the hypothermic elongated position should be maintained until the tissue cools (e.g., application of ice) (49, 69). Cryotherapy (application of cold) in conjunction with static (58) and proprioceptive neuromuscular facilitation (PNF) (20) stretching has demonstrated improved flexibility compared to stretching alone. Cryotherapy can induce anaesthetic effects which would allow the individual to push past their usual point of discomfort and stretch farther (increased stretch tolerance). However, as with almost every scientific question, there are dissenting findings (19, 57, 66). For example, there were no significant improvements in ROM with the application of ten minutes of cold water immersion (14) or ice packs (51). Although, hypothermic applications (cryotherapy) can aid in pain tolerance, it would also lead to vasoconstriction (14) and increase tissue visco-elasticity (69). Sapega et al. (69) suggests that cold should be used for therapeutic conditions when the objective is to disrupt adhesions or there is substantial muscle spasticity.

Stretching under Metastable Conditions

There are some innovative stretch coaches who have combined stretching routines under relatively unstable conditions. For example, Mario DiSanto of Argentina (a former national gymnastics champion) has gymnasts place one foot on the floor with another foot suspended from a Theraband (elastic band) attached to the ceiling. The athlete must stretch their muscles while maintaining stability and balance. A possible advantage of this type of active static stretching is the task specificity (11). Whereas, many stretching routines are performed under passive (no active voluntary muscles contraction) conditions, it is difficult to conceive of many sports that do not involve voluntary contractions and the necessity to maintain a high degree of balance or be in a state of strong metastability (39). Athletes and all individuals involved in activities of daily living continually move from states of stability (i.e., standing) to instability (movements such as walking and running) and return to stability again (i.e., return to the stance phase of walking or running). So, with a metastable state, an individual will move from a stable to a transiently unstable state (i.e., gymnast or figure skater or skier leaves the ground/floor to perform a jump or flip or other manoeuvre) and then return to their stable condition again. In many of these manoeuvres, a high degree of flexibility is necessary and the athlete certainly is not in a passive state. Stretching using unstable devices to create a metastable condition, accentuate task or action or sport specificity by placing the musculotendinous unit in a lengthened state under active metastable conditions. Thus, it is not only a stretch workout but a technique and motor control session as well (see Figure 7.1).

Figure 7.1 Stretching under unstable conditions to improve ROM and metastability

Summary

In summary, whereas single static stretches of five seconds can improve ROM, it is generally recommended that longer durations of 30–60 seconds provide optimal improvements in flexibility. It is not necessary to perform these stretches to the point of discomfort (100 percent intensity). While some studies have shown improvements even when stretching at 30–40 percent of maximal intensity, it seems that a stretching intensity of 60–85 percent would provide the greatest benefits. There are diurnal variations with flexibility and stretching midday to later in the day when the body is warmer decreases visco-elasticity and should enhance ROM. It is not recommended to stretch your back early in the morning when there is increased vertebral disc fluid pressure that could lead to disc protrusion injuries or nerve entrapment. There is no strong evidence for an optimal stretching frequency. However, since stretching is typically lower in intensity than resistance and high intensity anaerobic training without significant protein catabolism or tissue damage, it can be practised every day. Although, significant increases in flexibility are also experienced when stretching two or 3–six days per week.

The Konrad et al. meta-analysis (42) seems to be an irritating contrasting analysis since the lack of significant differences with their subgroup analyses of stretching volumes, intensities, and weekly frequencies contrasts with some of the original research articles. As I have told my students innumerable times: one study proves nothing! We need to look at the literature as a whole and that is the strength of meta-analyses. However meta-analyses are also not perfect. They attempt to congregate the findings from related but many times disparate groups and measures. For example, Konrad et al. (42) in their limitations section indicated that there was a moderate to high heterogeneity ($I^2 = 74.97$), owing to the wide variety of outcome measures, participants, and intervention durations. Thus, when you combine a wide distribution of findings, there will be higher variability and more difficulty in identifying anything significant. They also indicated the funnel plot as well as the significant Egger's regression intercept test revealed reporting bias. For example, significant positive results are more likely to be published in higher impact journals, producing greater citations numbers (24, 74). So, the good news is that it seems in general that many combinations of high or low stretching volumes, intensities or frequencies can be effective for improving ROM, whereas some original pieces of research suggest that for particular groups or scenarios there may be advantages with more strenuous or frequent stretching.

Whereas pre-event prolonged static stretching is not recommended, owing to the possibility of performance impairments, short to moderate duration static stretching (<60 seconds per /muscle group) within a full warm-up including dynamic stretching and dynamic activity does not impair performance (See Chapter 8). Post-activity stretching might involve a fatigued musculotendinous unit that could be susceptible to injury if subjected to intense elongation. Thus, post-activity stretching should be lower intensity, which can promote not only improvements in ROM but psychological feelings of relaxation. When stretching, breathing patterns should be at a slow and controlled rate with a full tidal volume (large breaths) to promote greater relaxation. The breath should be expired or held as you approach the end of the ROM. While stretching a warm muscle decreases visco-elasticity promoting a greater ease of ROM, the use of cryotherapy especially with intense stretching or rehabilitation may increase stretch (pain) tolerance allowing the individual to push farther than normal.

References

1. Alter, M.J. *Science of Flexibility.* Champaign, IL: Human Kinetics, 1996.
2. Apostolopoulos, N.Performance flexibility. In: *High-Performance Sports Conditioning*, B. Foran (Ed.). Champaign, IL: Human Kinetics, 2001, pp. 49–61.
3. Balaji, P.A., Varne, S.R., and Ali, S.S.Physiological effects of yogic practices and transcendental meditation in health and disease. *N Am J Med Sci* 4: 442–448, 2012.
4. Bandy, W.D. and Irion, J.M.The effect of time on the static stretch of the hamstrings muscles. *Physical Therapy* 74: 845–850, 1994.
5. Bandy, W.D., Irion, J.M., and Briggler, M.The effect of time and frequency of static stretching on flexibility of the hamstring muscles. *Physical Therapy* 77: 1090–1096, 1997.
6. Behm, D.G. Bambury, A., Cahill, F., and Power, K.Effect of acute static stretching on force, balance, reaction time, and movement time. *Med Sci Sports Exerc* 36: 1397–1402, 2004.
7. Behm, D.G., Bradbury, E.E., Haynes, A.T., Hodder, J.N., Leonard, A.M., and Paddock, N.R.Flexibility is not related to stretch-induced deficits in force or power. *Journal of Sports Science and Medicine* 5: 33–42, 2006.
8. Behm, D.G., Button, D.C., and Butt, J.C.Factors affecting force loss with prolonged stretching. *Can J Appl Physiol* 26: 261–272, 2001.
9. Behm, D.G. and Kibele, A.Effects of differing intensities of static stretching on jump performance. *Eur J Appl Physiol* 101: 587–594, 2007.
10. Behm, D.G., Plewe, S., Grage, P., Rabbani, A., Beigi, H.T., Byrne, J.M., and Button, D.C.Relative static stretch-induced impairments and dynamic stretch-induced enhancements are similar in young and middle-aged men. *Appl Physiol Nutr Metab* 36: 790–797, 2011.
11. Behm, D.G. and Sale, D.G.Velocity specificity of resistance training. *Sports Med* 15: 374–388, 1993.
12. Ben, M. and Harvey, L.A.Regular stretch does not increase muscle extensibility: a randomized controlled trial. *Scand J Med Sci Sports*, 2009.
13. Borsa, P.A., Sauers, E.L., and Herling, D.E.Patterns of glenohumeral joint laxity and stiffness in healthy men and women. *Med Sci Sports Exerc* 32: 1685–1690, 2000.
14. Burke, D.G., Holt, L.E., Rasmussen, R., MacKinnon, N.C., Vossen, J.F., and Pelham, T.W.Effects of Hot or Cold Water Immersion and Modified Proprioceptive Neuromuscular Facilitation Flexibility Exercise on Hamstring Length. *J Athl Train* 36: 16–19, 2001.
15. Campbell, E.J.M.Accessory muscles. In: *The respiratory muscles mechanics and neural control.* Philadelphia, PA: W.B. Saunders, 1970, pp. 181–193.

16. Carpenter, M.G., Frank, J.S., and Silcher, C.P.Surface height effects on postural control: a hypothesis for a stiffness strategy for stance. *J Vestib Res* 9: 277–286, 1999.
17. Chan, S.P., Hong, Y., and Robinson, P.D.Flexibility and passive resistance of the hamstrings of young adults using two different static stretching protocols. *Scandinavian Journal of Medicine and Science* 11: 81–86, 2001.
18. Chtourou, H., Aloui, A., Hammouda, O., Chaouachi, A., Chamari, K., and Souissi, N.Effect of static and dynamic stretching on the diurnal variations of jump performance in soccer players. *PLoS One* 8: e70534, 2013.
19. Cornelius, W.L.Exercise beneficial to the hip but questionable for the knee *NSCA Journal* 5: 40–41, 1984.
20. Cornelius, W.L.J.*et al.*The effects of cryotherapy and PNF on hip extensor flexibility. *Athletic Training* 19: 183–199, 1984.
21. Coulter, D. *Anatomy of batha yoga*. Honesdale, PA: Body and Breath, 2001.
22. Dane, S., Caliskan, E., Karasen, M., and Oztasan, N.Effects of unilateral nostril breathing on blood pressure and heart rate in right-handed healthy subjects. *Int J Neurosci* 112: 97–102, 2002.
23. Davis, D.S., Ashby, P.E., McHale, K.L., McQuain, J.A., and Wine, J.M.The effectiveness of 3 stretching techniques on hamstring flexibility using consistent stretching parameters. *Journal of Strength and Conditioning Research* 19: 27–32, 2005.
24. Easterbrook, P.J., Berlin, J.A., Gopalan, R., and Matthews, D.R.Publication bias in clinical research. *Lancet* 337: 867–872, 1991.
25. Farinatti, P.T., Brandao, C., Soares, P.P., and Duarte, A.F.Acute effects of stretching exercise on the heart rate variability in subjects with low flexibility levels. *J Strength Cond Res* 25: 1579–1585, 2011.
26. Fasen, J.M., O'Connor, A.M., Schwartz, S.L., Watson, J.O., Plastaras, C.T., Garvan, C.W., Bulcao, C., Johnson, S.C., and Akuthota, V.A randomized controlled trial of hamstring stretching: comparison of four techniques. *J Strength Cond Res* 23: 660–667, 2009.
27. Feland, J.B., Myrer, J.W., Schulthies, S.S., Fellingham, G.W., and Measom, G.W. The effect of duration of stretching of the hamstring muscle group for increasing range of motion in people aged 65 years or older. *Phys Ther* 81: 1110–1117, 2001.
28. Fleck, S.J. and Kraemer, W.J.Resistance Training: physiological responses and adaptations (Part 2 of 4). *The Physician and Sportsmedicine* 16: 108–118, 1988.
29. Fleck, S.J. and Kraemar, W.J.Resistance training: Physiological responses and adaptations (Part 3 of 4). *The Physician and Sports Medicine* 16: 63–72, 1988.
30. Fleck, S.J. and Kraemer, W.J.Resistance training: Basic principles (Part 1 of 4). *The Physician and Sports Medicine* 16: 108–114, 1988.
31. Freitas, S.R., Mendes, B., Le Sant, G., Andrade, R.J., Nordez, A., and Milanovic, Z.Can chronic stretching change the muscle-tendon mechanical properties? A review. *Scand J Med Sci Sports*, 2017.
32. Gonzalez, A.M., Hoffman, J.R., Townsend, J.R., Jajtner, A.R., Boone, C.H., Beyer, K.S., Baker, K.M., Wells, A.J., Mangine, G.T., Robinson, E.H., Church, D. D., Oliveira, L.P., Willoughby, D.S., Fukuda, D.H., and Stout, J.R.Intramuscular anabolic signaling and endocrine response following high volume and high intensity resistance exercise protocols in trained men. *Physiol Rep* 3, 2015.
33. Grabow, L., Young, J.D., Alcock, L.R., Quigley, P.J., Byrne, J.M., Granacher, U., Skarabot, J., and Behm, D.G.Higher Quadriceps Roller Massage Forces Do Not Amplify Range-of-Motion Increases nor Impair Strength and Jump Performance. *J Strength Cond Res* 32: 3059–3069, 2018.

34. Hamilton, A.R., Beck, K.L., Kaulbach, J., Kenny, M., Basset, F.A., DiSanto, M. C., and Behm, D.G.Breathing Techniques Affect Female but Not Male Hip Flexion Range of Motion. *J Strength Cond Res* 29: 3197–3205, 2015.
35. Hardy, M. and Woodall, W.Therapeutic effects of heat, cold, and stretch on connective tissue. *J Hand Ther* 11: 148–156, 1998.
36. Henricson, A.S., Fredriksson, K., Persson, I., Pereira, R., Rostedt, Y., and Westlin, N.E.The effect of heat and stretching on the range of hip motion*. *J Orthop Sports Phys Ther* 6: 110–115, 1984.
37. Huang, S.Y., Di Santo, M., Wadden, K.P., Cappa, D.F., Alkanani, T., and Behm, D.G.Short-duration massage at the hamstrings musculotendinous junction induces greater range of motion. *J Strength Cond Res* 24: 1917–1924, 2010.
38. Inami, T., Shimizu, T., Baba, R., and Nakagaki, A.Acute changes in autonomic nerve activity during passive static stretching. *American Journal of Sports Science and Medicine* 2: 166–170, 2014.
39. Kibele, A., Granacher, U., Muehlbauer, T., and Behm, D.G.Stable, Unstable and Metastable States of Equilibrium: Definitions and Applications to Human Movement. *J Sports Sci Med* 14: 885–887, 2015.
40. Knudson, D., Bennett, K., Corn, R, Leick, D., and Smith, C.Acute effects of stretching are not evident in the kinematics of the vertical jump. *Journal of Strength and Conditioning Research* 15: 98–101, 2001.
41. Knudson, D.V., Noffal, G.J., Bahamonde, R.E., Bauer, J.A., and Blackwell, J.R. Stretching has no effect on tennis serve performance. *J Strength Cond Res* 18: 654–656, 2004.
42. Konrad, A., Alizadeh, S., Daneshjoo, A., Anvar, S.H., Graham, A., Zahiri, A., Goudini, R., Edwards, C., Scharf, C., and Behm, D.G.Chronic effects of stretching on range of motion with consideration of potential moderating variables: A systematic review with meta-analysis. *J Sport Health Sci*, 2023.
43. Kraemer, W.J., Fleck, S., and Evans, W.J.Strength and Power Training: Physiological Mechanisms of Adaptation. In: *Resistance Training Adaptations.* 1990, pp. 363–397.
44. Kraemer, W.J. and Fleck, S.Resistance training: Exercise Prescription. *Physician and Sports Medicine* 16: 69–81, 1988.
45. Kraemer, W.J. and Ratamess, N.A.Fundamentals of resistance training: progression and exercise prescription. *Med Sci Sports Exerc* 36: 674–688, 2004.
46. Kubo, K., Kanehisa, H., and Fukunaga, T.Effect of stretching training on the viscoelastic properties of human tendon structures in vivo. *Journal Applied Physiology* 92: 595–601, 2002.
47. Kubo, K., Kanehisa, H., and Fukunaga, T.Effects of transient muscle contractions and stretching on the tendon structures in vivo. *Acta Physiol Scand* 175: 157–164, 2002.
48. Laban, M.M.Collagen tissue: implications of its response to stress in vitro. *Arch Phys Med Rehabil* 43: 461–466, 1962.
49. Lehmann, J.F., Masock, A.J., Warren, C.G., and Koblanski, J.N.Effect of therapeutic temperatures on tendon extensibility. *Arch Phys Med Rehabil* 51: 481–487, 1970.
50. Lehmann, J.F.*et al.*Diathermy and superficial heat, laser, and cold therapy. In: *Krusen's handbook of physical medicine and rehabilitation*. Philadelphia, PA: Saunders. 1990, pp. 285–367.

51. Lentell, G., Hetherington, T., Eagan, J., and Morgan, M.The use of thermal agents to influence the effectiveness of a low-load prolonged stretch. *J Orthop Sports Phys Ther* 16: 200–207, 1992.
52. Mahieu, N.N., McNair, P., De Muynck, M., Stevens, V., Blanckaert, I., Smits, N., and Witvrouw, E.Effect of static and ballistic stretching on the muscle-tendon tissue properties. *Med Sci Sports Exerc* 39: 494–501, 2007.
53. Manoel, M.E., Harris-Love, M.O., Danoff, J.V., and Miller, T.A.Acute effects of static, dynamic, and proprioceptive neuromuscular facilitation stretching on muscle power in women. *JStrength CondRes* 22: 1528–1534, 2008.
54. Mason TR, B.J.Thermal transition in collagen. *Biochemica et Biophysica Acta* PN 1254: 448–450, 1963.
55. McGill, S.M. *Low back disorders: Evidence based prevention and rehabilitation.* Champaign, IL: Huma Kinetics Publishers, 2002.
56. McHugh, M.P., Kremenic, I.J., Fox, M.B., and Gleim, G.W.The role of mechanical and neural restraints to joint range of motion during passive stretch. *MedSci-Sports Exerc* 30: 928–932, 1998.
57. Minton J.Student Writing Contest-1st Runner-up: A Comparison of Thermotherapy and Cryotherapy in Enhancing Supine, Extended-leg, Hip Flexion. *J Athl Train* 28: 172–176, 1993.
58. Newton, R.A.Effects of vapocoolants on passive hip flexion in healthy subjects. *Phys Ther* 65: 1034–1036, 1985.
59. Noonan, T.J., Best, T.M., Seaber, A.V., and Garrett, W.E., Jr. Thermal effects on skeletal muscle tensile behavior. *Am J Sports Med* 21: 517–522, 1993.
60. Peres, S., Draper, D.O., Knight, K.L., and Richard, M.D.Pulsed shortwave diathermy and prolonged stretch increase dorsiflexion range of motion more than prolonged stretch alone. *Journal of Athletic Training* 2: S49, 2001.
61. Rancour, J., Holmes, C.F., and Cipriani, D.J.The effects of intermittent stretching following a 4-week static stretching protocol: a randomized trial. *JStrength CondRes* 23: 2217–2222, 2009.
62. Rigby, B.J.The Effect of Mechanical Extension Upon the Thermal Stability of Collagen. *Biochim Biophys Acta* 79: 634–636, 1964.
63. Rigby, B.J., Hirai, N., Spikes, J.D., and Eyring, H.The Mechanical Properties of Rat Tail Tendon. *J Gen Physiol* 43: 265–283, 1959.
64. Roaf, R. *Posture.* New York: Academic Press, 1977.
65. Roberts, J.M. and Wilson, K.Effect of stretching duration on active and passive range of motion in the lower extremity. *BrJSports Med* 33: 259–263, 1999.
66. Rosenberg, B.S., Cornelius, W.L., and Jackson, A.W.The effects of cryotherapy and PNF stretching techniques on hip extensor flexibility in elderly females. *Journal of Physical Education and Sport Science* 2: 31–36, 1990.
67. Ross, M.D.Effect of a 15-day pragmatic hamstring stretching program on hamstring flexibility and single hop for distance test performance. *ResSports Med* 15: 271–281, 2007.
68. Ryan, E.D., Herda, T.J., Costa, P.B., Defreitas, J.M., Beck, T.W., Stout, J., and Cramer, J.T.Determining the minimum number of passive stretches necessary to alter musculotendinous stiffness. *J Sports Sci* 27: 957–961, 2009.
69. Sapega, A.A., Quedenfeld, T.C., Moyer, R.A., and Butler, R.A.Biophysical Factors in Range-of-Motion Exercise. *Phys Sportsmed* 9: 57–65, 1981.

70. Sawyer, P.C., Uhl, T.L., Mattacola, C.G., Johnson, D.L., and Yates, J.W.Effects of moist heat on hamstring flexibility and muscle temperature. *J Strength Cond Res* 17: 285–290, 2003.
71. Seals, D.R., Suwarno, N.O., and Dempsey, J.A.Influence of lung volume on sympathetic nerve discharge in normal humans. *Circ Res* 67: 130–141, 1990.
72. Sechrist, W.C. and Stull, G.A.Effect of mild activity, heat applications, and cold applications on range of joint movement. *Am Correct Ther J* 23: 120–123, 1969.
73. Sun, J.S., Tsuang, Y.H., Liu, T.K., Hang, Y.S., Cheng, C.K., and Lee, W.W.Viscoplasticity of rabbit skeletal muscle under dynamic cyclic loading. *Clin Biomech (Bristol, Avon)* 10: 258–262, 1995.
74. Sutton, A.J., Duval, S.J., Tweedie, R.L., Abrams, K.R., and Jones, D.R.Empirical assessment of effect of publication bias on meta-analyses. *BMJ* 320: 1574–1577, 2000.
75. Taylor, D.C., Dalton, J.D., Seaber, A.V., and Garret, W.E.Viscoelastic properties of muscle-tendon units: the biomechanical effects of stretching. *The American Journal of Sports Medicine* 18: 300–308, 1990.
76. Thomas, E., Bianco, A., Paoli, A., and Palma, A.The Relation Between Stretching Typology and Stretching Duration: The Effects on Range of Motion. *Int J Sports Med* 39: 243–254, 2018.
77. Trajano, G.S., Nosaka, K., Seitz, L.B., and Blazevich, A.J.Intermittent stretch reduces force and central drive more than continuous stretch. *Med Sci Sports Exerc* 46: 902–910, 2014.
78. Volek, J., Kraemer, W., Bush, J., Incledon, T., and Boetes, M.Testosterone and cortisol in relationship to dietary nutrients and resistance exercise. *J Appl Physiol* 82: 49–54, 1997.
79. Wallin, D., Ekblom, B., Grahn, R., and Nordenborg, T.Improvement of muscle flexibility. A comparison between two techniques. *Am J Sports Med* 13: 263–268, 1985.
80. Walter, J., Figoni, S.F., Andres, F.F., and Brown, E.Training intensity and duration in flexibility. *Clinical Kinesiology* 2: 40–45, 1996.
81. Warren, C.G., Lehmann, J.F., and Koblanski, J.N.Elongation of rat tail tendon: effect of load and temperature. *Arch Phys Med Rehabil* 52: 465–474 passim, 1971.
82. Warren, C.G., Lehmann, J.F., and Koblanski, J.N.Heat and stretch procedures: an evaluation using rat tail tendon. *Arch Phys Med Rehabil* 57: 122–126, 1976.
83. Wessling, K.C., DeVane, D.A., and Hylton, C.R.Effects of static stretch versus static stretch and ultrasound combined on triceps surae muscle extensibility in healthy women. *Phys Ther* 67: 674–679, 1987.
84. Ylinen, J., Kankainen, T., Kautiainen, H., Rezasoltani, A., Kuukkanen, T., and Hakkinen, A.Effect of stretching on hamstring muscle compliance. *J Rehabil Med* 41: 80–84, 2009.
85. Young, W., Elias, G., and Power, J.Effects of static stretching volume and intensity on plantar flexor explosive force production and range of motion. *Journal of Sports Medicine and Physical Fitness* 46: 403–411, 2006.
86. Yuktasir, B. and Kaya, F.Investigation into the long term effects of static and PNF stretching exercises on range of motion and jump performance. *Journal of Bodywork and Movement Therapies* 13: 11–21, 2009.

8 Stretching Effects on Injury Reduction and Health

The triumvirate of reasons for stretching over the decades has been to increase range of motion (ROM), improve performance and decrease the chances for injury. The research is overwhelming that we can increase flexibility with stretching (55, 56). However, the next two traditional reasons for stretching are less clear. We will discuss in Chapter 8 how an acute session of prolonged static stretching might actually impair performance rather than improve it. The contention that stretching reduces injuries is also mired in controversy.

An Australian researcher Rodney Pope is well cited for his work in the area. Pope et al. (69) investigated 1,093 Australian army recruits during 12 weeks of military training and found a significant correlation between dorsiflexion range of motion (ROM) and injury incidence. The injuries they followed were ankle sprains, tibia or foot stress fractures, tibial periostitis, Achilles tendonitis, and anterior tibial component syndrome. Poor flexibility was associated with 2.5 times the risk for such injuries compared to average dorsiflexion ROM and eight times the risk compared to people with a high level of flexibility. However, the stretch training programme had no significant effect on the incidence of these injuries. Do these two statements not conflict with each other? Not entirely! The amount of flexibility is an intrinsic factor (something that you have or don't have), whereas stretching is an extrinsic factor (something that you do). So, in the first Pope study, in general if you had better dorsiflexion ROM then you were less likely to get injured. However, doing a static stretching programme did not change the risks to a significant degree. In a similar later study, Pope et al. (70) involved 1,538 Australian army recruits who performed one static stretch each for six lower leg muscles every second day. The stretch training programme did not produce a clinically worthwhile reduction in lower limb injury risk. Injuries included lower body stress fractures, muscle strains, ligament sprains, periostitis, tendonitis, meniscal lesions, compartment syndromes, and bursitis among others. Thus, neither of Pope's studies showed an extrinsic effect of stretching on injury incidence. Small and colleagues (78) reported similar results in their review. Four randomized clinical trials found that static stretching was ineffective in reducing the incidence of exercise-related injury. One of three controlled clinical trials indicated that static stretching decreased

DOI: 10.4324/9781032709086-9

the incidence of exercise-related injury. However, three of the seven studies reported reductions in musculotendinous and ligament injuries although there were no statistically significant reductions in the all types of injury risk. Small concluded that there was moderate to strong evidence that static stretching does not reduce overall injury rates, but may reduce musculotendinous injuries. Weldon and Hill (87) said they could not make a definitive statement, owing to the paucity of well controlled studies at that time. They went as far to say that the scant evidence prior to 2003 would suggest that pre-exercise stretching may actually increase the risk of injury. However, they did not seem that convinced as they also offered that scientific and clinical evidence indicated that stretching in the post-exercise period (cool down) should increase the energy absorbing capabilities of the musculotendinous unit (MTU) reducing the risk of injury. Shrier (77) in his review also indicated that stretching before an event will not reduce injuries. An interesting retrospective study of 3,669 runners found that as you would expect that there was an increased chance of injury in runners if they weighed more, had a previous injury, competed in races (racing at higher intensities and speeds), trained more than two hours and/or more than 20 km per week (higher volumes of training), but surprisingly also if they stretched before running (75). One might speculate that previously injured runners might stretch more to prevent further injuries and that is why stretching was associated with injuries. However, stretching before running was independent of injury history. Based on the hypothesis that stretch training can increase stretch and pain tolerance (62), the runners, if stretching to maximum intensity or point of maximum discomfort may elongate their muscles or tendons past a safe limit and unknowingly inflict muscle and connective tissue injuries (87). In contrast, a randomized, single-blind, non-supervised (self-reports), controlled trial, with self-reporting of muscle, ligament, and tendon injuries found that distance runners who stretched pre- and post-running for 12 weeks experienced a reduction in injuries (0.66 injuries per person-year with stretching versus 0.88 injuries per person-year with controls) (52). Just to confuse the public a bit more, a review by Woods (93) contradicted the earlier reviews by stating that using particular techniques or protocols would provide a reduction in injuries. They indicated that by elongating muscle length to provide a greater ROM led to less musculotendinous unit injuries. They recommended that it was important to ensure an anterior pelvic tilt with the hip at 90° of flexion when stretching the hamstrings. Similarly, a 12-week stretching programme with male youth soccer players, reported increased ROM and a reduction in muscle tightness, which may have contributed to the reduction in non-contact lower limb, trunk, muscle and tendon injuries (6).

The confusion resides in whether the reviews are examining "all-cause" injury risk versus musculotendinous unit injury risk. For example, Amako et al. (4) evaluated military recruits who performed static stretching for 3-months reporting no effect on all-cause injury risk but a significant reduction in musculotendinous injuries, as well as low back pain. How can pre-event stretching be expected to reduce lower body fractures, meniscal lesions, bursitis or other afflictions? Will stretching reduce overtraining effects typical of long-distance runners and cyclists? It is unlikely according to the literature. But if stretching can improve energy absorbing capabilities of the musculotendinous unit leading to better running economy (lower aerobic output for a given running workload) (38), then you might expect a positive trend for stretching to prevent musculotendinous injuries (66, 90), owing especially to high loads associated with rapid force absorption such as sprinting and agility. Our 2021 review supported that both acute and chronic stretching can reduce the incidence of musculotendinous injuries especially in running-based sports and with explosive actions and change of direction (14). We suggested that this musculotendinous injury protection could be related to stretch-induced changes in the force-length relationship (increased force available at longer muscle lengths) as well as reduced active musculotendinous stiffness, among other factors. As many muscle and tendon injuries occur at extended muscle lengths, greater force at this length could have a protective function. It was also discovered that while the acute effects of stretching on balance is equivocal, chronic stretch training can improve balance, which may contribute to a decreased incidence of injuries associated of slips, trips, and falls (14).

In another one of our reviews (12), Dr. Malachy McHugh (a proud Irishman working in New York) was responsible for the section on injury incidence. In his review of stretching articles investigating injury evidence, he found that eight studies showed some effectiveness of stretching, whereas four showed no effect. There was no evidence that stretching negatively influences (increases) injury risk. Only two of five studies that stretched for less than five minutes indicated reduced injury risk whereas five of six stretching studies with more than five minutes showed an injury incidence benefit for stretching. Only two of five stretching studies showed injury reduction benefits for endurance sports or military training which typically involve overuse injuries. Five of 6 studies showed fewer injuries with sprint running type of sports. The effectiveness of stretching for sprint activities could be related to the high tensile demands placed on the musculotendinous unit with the stretch-shortening cycle. Witvrouw et al. (91) suggest that by increasing the compliance of the tendon, the tissue will have greater energy absorbing capacity. However, as static stretching may not be as effective as ballistic stretching for decreasing tendon stiffness, they recommend including ballistic stretching within the warm-up. Fradkin (36) in another review reported that three of five studies they reviewed showed significant injury prevention benefits with little evidence to suggest stretching was harmful in terms of inflicting injuries. Furthermore, most research (except for one study [22]) reports that stretching does not reduce muscle soreness or other symptoms of muscle damage. Stretching also does not improve recovery from fatigue and may actually inhibit recovery (32). Although, it is a very common practice, one article reports that acute static and proprioceptive neuromuscular facilitation (PNF) stretching did not prevent the frequency of muscle cramping (68). Thus, in summary, pre-activity stretching of five minutes or more should be more beneficial for injury prevention with sprint running but less effective with endurance type running activities (9)(including military training) that tend to lead to overuse injuries. As workplace injuries are also often overuse injuries, the recommendation to stretch at work may not be as effective as it is often promoted. Behm et al. (12) report the studies in their review indicate a 54 percent risk reduction in acute muscle and tendon injuries associated with stretching.

But, Chapter 8 will emphasize to the reader that more than 60 seconds of static stretching per muscle group without a full warm-up could lead to performance impairments. Important to note that the research says "per muscle group". To reduce the risk of injury incidence you should stretch more than one muscle group. So, to reduce lower body injuries you might want to static stretch the hip flexors and extensors, knee extensors and flexors, abductors and adductors, dorsiflexors, plantar flexors and ankle evertors for 60 seconds less "each" resulting in perhaps nine minutes of lower body stretching.

Physiological Rationale for Stretch-induced Injury Reduction

Why would stretching have an effect on injury incidence especially with higher speed running activities like sprinting or change of direction (e.g., agility)? Unfortunately, most of the studies that investigate injury risk have not provided a rationale for reduced injuries (66). Theoretically, it is possible that stretching makes the musculotendinous unit more compliant (67, 84), increased compliance shifts the angle–torque relationship to allow greater relative force production at longer muscle lengths (48, 67), and thus the improved resistance to excessive muscle elongation may decrease the susceptibility to muscle strain. Similarly, stretching for 5 minutes can increase tendon compliance and decrease hysteresis (58), which would also allow the tendon to absorb forces over a more extended duration decreasing the stress on this connective tissue minimizing the possibility of a strain. However, a competing rationale could be that increased muscle force output from a lengthened muscle could increase the likelihood of injury. But this rationale would not apply to the risk of other injuries such as ligament sprains, skeletal fractures or overuse injuries. In general, the ability of the MTU to better absorb force perturbations whether it is due to less MTU stiffness (greater compliance) or higher force capabilities at longer muscle lengths should reduce injuries especially with high forces and torques associated with sprinting and agility. Dr. McHugh provided the following theory and application in a personal correspondence. "Muscle strains can occur during concentric contractions and can occur at short muscle lengths. For example, in the days when soccer balls were made with leather and the ball became much heavier in wet conditions it was common to see rectus femoris strains in players kicking a ball (often complete ruptures). At the point of impact with the ball the rectus femoris MTU is shortening (knee is extending, hip is flexing). However, the brain has programmed the body to expect impact with an object with a particular mass.

The actual mass is in fact greater than expected, owing to the wet conditions. At the point of impact with the ball, the contractile and non-contractile elements are shortening synchronously to optimize the impulse imparted on the ball. When the resistance to movement provided by the ball is greater than expected this subtly changes the internal muscle tendon mechanics; at the knee quadriceps and patellar tendon recoil act to extend the knee, if the rate of knee extension is slowed by a heavier than expected ball this will change the loading at the muscle tendon junction (in simplistic terms the tendon is now pulling on the muscle fibres mores than the bone because an external force has slowed joint movement). I think that anything that disrupts the interaction between tendon and muscle mechanics during forceful stretch-shortening cycle movements can cause injury to the muscle-tendon unit."

Dr. McHugh continued as follows: "I think the ability of a contracted muscle to absorb sudden tensile loads is important in the mechanism of muscle strains. If an isometrically contracted muscle is stretched suddenly (even a little stretch) there is a large increase in force that is due to the mechanical resistance of the MTU (65, 86). In many cases, owing to the very brief onset, the stretch reflex or other neural changes cannot contribute. I think if the contractile component is able to resist the elongation and impart it to the tendon the muscle will be protected from potential strain injury. Stretching might impact this (albeit in a small way) by allowing greater myofilament overlap at longer muscle lengths than was achievable before the acute bout of stretching, thus allowing the muscle tendon unit elongation to be transferred to the non-contractile component."

While greater compliance will play a role in shock and stress absorption, a musculotendinous unit that is too slack will be hampered in its ability to effectively transfer force from the muscle to the bone to initiate movement. Paradoxically, the muscles and tendons must be compliant and relatively stiff at the same time to prevent injuries while also optimizing performance respectively (14). According to Thorpe et al. (82, 83), the tendon stiffness to muscle strength ratio of individual tendon fibres must be high. However, whereas collagen has the constituency of copper (very stiff and strong), appropriate ROM and compliance can be achieved with efficient sliding between collagen fibres and fascicles. Thus, you can have active stiffness of individual muscle fibres and tendons but you can also have some compliance and an efficient ROM with filament sliding and fascicle rotation (12, 14).

Dynamic Stretching and Activity Effects on Injury Incidence

Our review on the effects of dynamic stretching on injury incidence (11) could only locate two studies that specifically examined dynamic stretching effects in isolation. One study showed no significant effect on injury incidence when comparing a dynamic stretching-only group and a combined dynamic plus static stretching group (95). The second article found increased ankle joint stability with functional dynamic stretching training with injured dancers

(96). However, there was abundant evidence revealing the positive effects of multi-faceted dynamic activity programmes incorporating both dynamic stretching and dynamic activity (i.e., FIFA 11+, FIFA 11 + Kids, FIFA 11 + S, HarmoKnee, Knäkontroll, SISU Idrottsböcker, NMT) (11). These programmes can involve dynamic stretching, jumping, running, bounding, agility, balance, and core stability. There are conflicting reports however on the effects of dynamic stretch training on strength, balance, proprioception, musculotendinous unit stiffness (11). A single session of dynamic stretching should promote thixotropic effects (reduced viscoelasticity owing to stress tension and contraction-induced increases in muscle temperature), contribute to a more positive emotional state, decrease muscle tension, as well as increase concentration, attention, and better prepare athletes for activity and competition (11). With the lack of evidence on dynamic stretching alone, more research is needed on this topic.

Stretching Effects on Posture, Lower Back Pain, and Compensatory Overuse Syndromes

A lack of flexibility in the pelvis, back or other joints can lead to poor posture (23). We strive to keep our centre of gravity over our base of support (equilibrium or balance) or we want to be able to temporarily move our centre of gravity outside our base of support and return to a balanced position (metastability) (54) when reaching or moving. Poor flexibility leading to poor posture as seen with rounded shoulders, or insufficient lumbar spinal curves can hamper our balance or movement capabilities. We then try to overcompensate by engaging other muscles and joints to achieve these functions. The shortened or inflexible muscles may also atrophy and become weaker leading to muscle imbalances (agonist to antagonist muscle activity or prime mover to synergist muscle activity). As the other muscles or joints may not be efficient at achieving these movements, they may be susceptible to fatigue, overuse and perhaps eventual injury. A combination of flexibility training, muscle strengthening and movement re-education might be needed to return the body to proper functioning and reduce the possibility of pain, discomfort and possible injury with these movements.

The role of improving back flexibility to reduce the incidence of low back pain is controversial. It seems that both ends of the spectrum: lack of back flexibility or mobility (19, 24, 85) and excessive spinal mobility (50) can contribute to back problems. However, others have reported that tight hip flexors (33), lack of lumbar or hip motion (33) or thoracic mobility (61) were not related to low back pain. Nonetheless, many authors emphasize the need for an adequate back or trunk mobility in order to ensure proper mechanical functioning (34) otherwise the excessive bending movements on individual vertebrae can lead to an increased injury risk (26). It is suggested by many researchers and organizations that adequate flexibility can reduce the incidence and severity of low back pain (5, 20, 25, 60, 71, 72). Based on the evidence for and against the protection of lower back problems with greater back ROM (7, 8, 64), Stu McGill suggests that trunk stretching should not emphasize maximum ROM, torso flexibility exercises should be limited to unloaded flexion and extension, ensure adequate hip and knee flexibility so that the back does not have to overcompensate (compensatory relative flexibility [73] for other inflexible joints and finally the first concern should be to improve the strength and endurance of the back prior to incorporating back or trunk flexibility programmes.

Effects of Immobilization on Flexibility

When an individual is injured and subjected to immobilization, the consequences can include muscle atrophy (31, 45, 46), decreases in dynamic (45) and static strength (88, 89), muscle activation (30, 37), reflex potentiation (74), and maximum motor unit firing rates (28, 29) as well as increases in the duration of twitch contractile properties (29, 88, 89). In addition, immobilization can adversely affect flexibility (10). Prolonged immobilization with the muscle in a shortened position can decrease the number of sarcomeres in series, shortening muscle fibre length (39, 79, 80). Fortunately, this sarcomere loss can be rapidly regenerated once normal activity and muscle lengthening occurs (39, 79, 80). Conversely, if a muscle is immobilized in a lengthened position, additional sarcomeres are added (79). Unfortunately, the biomechanics of our joints and muscles dictate that when one muscle is casted in a lengthened position, the other is typically in a shortened position and thus there is no strategy that can benefit both muscles at the same time.

Another factor contributing to a more inflexible muscle with immobilization would be a relative increase in connective tissue, as connective tissue degrades at a slower rate than muscle tissue contributing to a stiffer muscle (79, 80). In addition, there is reported to be a more acute angle of collagen to the axis of the muscle fibres compared to healthy muscle (40) that would also affect stiffness. Thirdly, immobilization reduces the water and glycosaminoglycan content of connective tissue decreasing the distance between collagen fibres resulting in abnormal cross-linkages that limit extensibility of the tissue (2), (63). Stretching tends to counter these deleterious effects (1). However, an immobilized limb in a cast may be difficult to stretch in order to prevent or reduce the decreased extensibility. A counter measure that may be incorporated while the muscle is still immobilized is to stretch the contralateral muscle or actually any other muscle. Unilateral stretching of the hamstrings has improved the ROM of the contralateral hip flexors (21). Furthermore, stretching either the shoulders or the adductor (groin) muscles resulted in improved ROM in the hip flexors and shoulders respectively (13). So, stretch your upper body and your lower body becomes more flexible or stretch your lower body and your upper body becomes more flexible. This is important information for individuals who are immobilized. Just like cross-education (strength increase in contralateral muscle after unilateral training)

(47, 49, 76, 97) is important to preserve strength in an immobilized muscle, stretching unaffected muscles should help to preserve flexibility or at least help to minimize loss of flexibility during immobilization.

Stretching Effects on Dysmenorrhea

Another painful or uncomfortable condition that may be alleviated by stretching is dysmenorrhea (3). Dysmenorrhea is greater than normal pain or discomfort association with menstruation. While it can be attributed in some cases to abnormal oestrogen-progesterone balance; poor posture, owing to shortening of uterine connective tissue which affects the pelvic posterior tilt has also been postulated (41, 71). Static stretching of the pelvis on a consistent basis has been reported to alleviate or reduce dysmenorrhea (41, 42, 43). It is speculated that the stretching decreases the stiffness or restrictions of the fascia and ligaments relieving the compression and irritation of the affected nerves (15, 16).

Stretching Effects on the Cardiovascular System

Stretching can have both positive and negative effects upon the cardiovascular system. One to four minutes of static stretching enhanced blood flow to the stretched muscle for 10 minutes after stretching (57). However, static stretching also acutely increases heart rate, blood pressure and rate pressure product: rate pressure product (RPP) = heart rate (HR) x systolic blood pressure (BP) − a measure of myocardial workload − especially when a Valsalva manoeuvre (exhaling against a closed glottis) was used during the stretching procedure. This increased cardiovascular strain could be dangerous for people with heart and cardiovascular problems (59). However, a meta-analysis revealed that chronic stretch training improved vascular endothelial function, reduced arterial stiffness, resting heart rate, and diastolic blood pressure in middle aged and older adults (53). Another meta-analysis provides further support for the benefits of acute and chronic stretching, reporting significant reductions in arterial stiffness and resting heart rate with no adverse effects on blood pressure with cardiovascular disease patients (81). An additional review indicates that stretch training can improve cardiac autonomic function (increased heart rate variability) in different populations, possibly owing to increases in baroreflex sensitivity, relaxation, and nitric oxide availability (92). Both acute dynamic and static stretching sessions provided large magnitude increases in oxygen saturation, with dynamic stretching showing greater magnitudes (☒ 63 percent) versus static (☒ 55 percent) stretching, which persisted for two minutes after stretching ceased (18). Furthermore, it has been shown that poor flexibility is associated with greater age-related arterial stiffening (94). Remember that associations and correlations are not causative. That means that improving your flexibility will not necessarily decrease arterial stiffness. It may just mean that people with stiff arteries have not been active throughout their lives and poor flexibility is just one sign that their sedentary lifestyle has contributed to possible arteriosclerosis. However, cyclical stretching can have positive effects on the cell shape, cytoplasmic organization and intracellular processes of vascular smooth muscle cells. When stretching, you are unlikely to substantially stretch the arteries but you can increase the vascular tonus with 1 minute of static stretching (51). As the vasculature cytoskeleton and contractile proteins are associated with and embedded into the extracellular matrix (ECM), stretching can move, deform or place stress on the surrounding ECM and thus stimulate the vascular cell membrane receptors (44). Twelve weeks of passive static stretching improved vascular function and arterial remodelling and arterial stiffness. The stretch-induced enhancements related to central mechanisms returned to baseline within six weeks of detraining (17). Repetitive stretching actions would induce phases of hypoemia and hyperemia, owing to repetitive changes in shearing stress and rate, which could be an effective stimulus to cause positive adaptations in vascular function, arterial stiffness and vascular remodeling (17). So, while we normally think of aerobic exercise and nutrition as being the most important factors for cardiovascular health, stretching can also contribute to a healthy cardiovascular system.

Stretching Effects on Diabetes

The day I was going to submit this second edition, I found this article and thought it should be included. A group from Brazil had diabetic participants (61.5 ± 6.6 years with type II diabetes for 9.4 years) perform eight different static stretches for either 30 or 45 seconds each over eight weeks (35). Both stretching durations showed significant reductions in acute and chronic glycemia. We already know that exercise and activity in general, as well as resistance training, are all beneficial in reducing the need for insulin. Folha et al. (35) demonstrated that the activity (static stretching) can be relatively passive and also reduce type II diabetes associated glycemic problems. How does this occur? My hypothesis is that even passive stretching will cause reflexive-induced contractions of the stretched muscle and probably involve active contractions of the target (agonist), synergist, and antagonist muscles to help move the joint to the full ROM and hold that position. When a muscle is contracted, IGF-1 (insulin-like growth factors) signals the GLUT-4 (glucose) transporters to translocate to the membrane to facilitate glucose entry into the muscle (27). Hence, if the membrane is less sensitive to insulin as with type II diabetes, the activation of GLUT-4 transporters would partially compensate for this insulin sensitivity deficit.

Summary

Generally, there is not strong evidence for stretching programmes to reduce the incidence of all cause injuries. However, there is more convincing, moderate levels of evidence that a static stretch training programme can reduce the incidence of musculotendinous and ligamentous injuries especially with more powerful actions such as sprinting and change of direction (agility) (12, 14) (Figure 8.1). Flexibility training may increase the energy absorbing capability of the musculotendinous unit, as a more compliant tissue will absorb these forces over a more prolonged duration. There is a lack of evidence regarding the effects of dynamic stretching alone on injury incidence but dynamic stretching combined with dynamic activity does seem to produce positive effects on injury prevention. It is unlikely that neural reflexes contribute greatly to this improved protection as the unexpected loads or torques from a slip, fall or collision may be too rapid to be affected by a reflex response. Chronic problems such as low back pain might be positively affected by stretch training as the increased flexibility may improve hip/pelvic orientation affecting spinal curves alleviating back pain and compensatory overuse syndromes. Similarly, stretching may decrease the discomfort of dysmenorrhea by reducing fascia and ligament stiffness and restrictions. It is important to stretch an immobilized limb as there can be a shortening of the muscle, owing primarily to abnormal cross linkages and in very prolonged cases decreases in sarcomeres in series. Even stretching another muscle such as the contralateral homologous or heterologous muscle can help maintain a certain degree of ROM in the affected muscle. While enhanced flexibility is correlated with decreased arterial stiffness and stretching can increase blood

Injury Risk

Little evidence that SS decreases all-cause injury risk.
Stronger evidence for a SS-induced reduction in MTU injuries, particularly in running-based sports.
Ballistic stretching and active warm-up speculated to reduce tendon injuries in SSC activities

<u>Speculative Mechanisms</u>
Altered muscle force length relationship
More compliant MTU
Develop appropriate tendon stiffness to muscle strength ratio
Improvements in balance

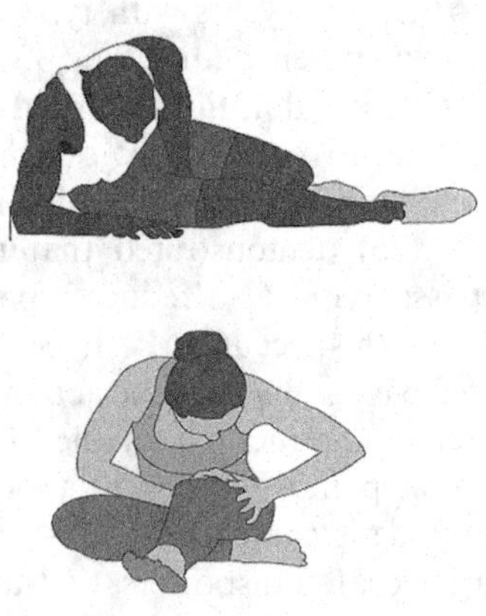

Figure 8.1 Injury risk factors

flow to the stretched muscle, individuals with cardiovascular diseases must be cautious, as a result of stretch-induced increases in the rate pressure product (heart rate x blood pressure: a measure of cardiac workload or demand).

References

1. Ahtikoski, A.M., Koskinen, S.O., Virtanen, P., Kovanen, V., and Takala, T.E.Regulation of synthesis of fibrillar collagens in rat skeletal muscle during immobilization in shortened and lengthened positions. *Acta Physiol Scand* 172: 131–140, 2001.
2. Akeson, W.H., Amiel, D., and Woo, S.Immobility effects on the synovial joints: The pathomechnics of joint contraction. *Biorheology* 1/2: 95–110, 1980.
3. Alter, M.J. *Science of Flexibility.* Champaign, IL: Human Kinetics, 1996.
4. Amako, M., Oda, T., Masuoka, K., Yokoi, H., and Campisi, P.Effect of static stretching on prevention of injuries for military recruits. *Mil Med* 168: 442–446, 2003.
5. American College of Sports Medicine. ACSM's Guidelines for Exercise Testing and Prescription. *Medicine Science Sports and Exercise* 6: 158–164, 2000.
6. Azuma, N. and Someya, F.Injury prevention effects of stretching exercise intervention by physical therapists in male high school soccer players. *Scand J Med Sci Sports* 30: 2178–2192, 2020.
7. Battie, M.C., Biggos, S.J., Fisher, L.D., Spengler, D.M., Hansson, T.H., Nachemson, A.L., and Wortley, M.D.The role of spinal flexibility in back pain complaints within industry: A prospective study. *Spine* 8: 768–773, 1990.
8. Battie, M.C., Biggos, S.J., Sheely, A., and Wortley, M.D.Spinal flexibility and individual factors that influence it. *Physical Therapy*: 67, 1987.
9. Baxter, C., McNaughton, L.R., Sparks, A., Norton, L., and Bentley, D.Impact of stretching on the performance and injury risk of long-distance runners. *Res Sports Med* 25: 78–90, 2017.
10. Behm, D.G.Debilitation to adaptation. *Journal of Strength and Conditioning Research* 7: 65–75, 1993.
11. Behm, D.G., Alizadeh, S., Daneshjoo, A, and Konrad, A.Potential Effects of Dynamic Stretching on Injury Incidence of Athletes: A Narrative Review of Risk Factors. *Sports Med* 53: 1359–1373, 2023.
12. Behm, D.G., Blazevich, A.J., Kay, A.D., and McHugh, M.Acute effects of muscle stretching on physical performance, range of motion, and injury incidence in healthy active individuals: a systematic review. *Appl Physiol Nutr Metab* 41: 1–11, 2016.
13. Behm, D.G., Cavanaugh, T., Quigley, P., Reid, J.C., Nardi, P.S., and Marchetti, P. H.Acute bouts of upper and lower body static and dynamic stretching increase nonlocal joint range of motion. *Eur J Appl Physiol* 116: 241–249, 2016.
14. Behm, D.G., Kay, A.D., Trajano, G.S., Alizadeh, S., and Blazevich, A.J.Effects of stretching on injury risk reduction and balance. *Journal of Clinical Exercise Physiology* 10: 106–116, 2021.
15. Billing, H.E.Dysmenorrhea: The result of a postural defect. *Archive of Surgery* 5: 611–613, 1943.
16. Billing, H.E.Fascial Stretching. *Journal of Physical and Mental Rehabilitation* 1: 4–8, 1951.
17. Bisconti, A.V., Ce, E., Longo, S., Venturelli, M., Coratella, G., Limonta, E., Doria, C., Rampichini, S., and Esposito, F.Evidence for improved systemic and

local vascular function after long-term passive static stretching training of the musculoskeletal system. *J Physiol* 598: 3645–3666, 2020.
18. Brodeur, Z.R., Paustian, M.J., Monteleone-Haught, D.A., Lamm, R.A., Pagano, A.G., and Ellis, C.E.The Effects of Static and Dynamic Stretching on Muscle Oxygen Saturation in the Rectus Femoris. *Int J Exerc Sci* 15: 702–708, 2022.
19. Burton, A.K., Tillotson, K.M., and Troup, J.D.Variation in lumbar sagital mobility with low-back. *Spine* 6: 584–590, 1989.
20. Cailliet, R. *Low back pain syndrome.* Philadelphia, PA: F.A. Davis, 1988.
21. Chaouachi, A., Padulo, J., Kasmi, S., Othmen, A.B., Chatra, M., and Behm, D.G. Unilateral static and dynamic hamstrings stretching increases contralateral hip flexion range of motion. *Clin Physiol Funct Imaging* 37: 23–29, 2017.
22. Chen, H.M., Wang, H.H., Chen, C.H., and Hu, H.M.Effectiveness of a stretching exercise program on low back pain and exercise self-efficacy among nurses in Taiwan: a randomized clinical trial. *Pain Manag Nurs* 15: 283–291, 2014.
23. Corbin, C.B. and Noble, L.Flexibility: A major component of physical fitness. *Journal of physical Education and Recreation* 6: 23–24, 57–60, 1980.
24. Curtis, L., Mayer, T.G., and Gatchel, R.J.Physical progress and residual impairment quantification after functional restoration. Part III: Isokinetic and isoinertial lifting capacity. *Spine (Phila Pa 1976)* 19: 401–405, 1994.
25. Deyo, R.A., Walsh, N.E., Martin, D.C., Schoenfeld, L.S., and Ramamurthy, S.A controlled trial of transcutaneous electrical nerve stimulation (TENS) and exercise for chronic low back pain. *N Engl J Med* 322: 1627–1634, 1990.
26. Dolan, P. and Adams, M.A.Influence of lumbar and hip mobility on the bending stresses acting on the lumbar spine. *Clin Biomech (Bristol, Avon)* 8: 185–192, 1993.
27. Douen, A.G., Ramlal, T., Klip, A., Young, D.A., Cartee, G.D., and Holloszy, J.O. Exercise-induced increase in glucose transporters in plasma membranes of rat skeletal muscle. *Endocrinology* 124: 449–454, 1989.
28. Duchateau, J. and Hainaut, K.Electrical and mechanical changes in immobilized human muscle. *Journal of Applied Physiology* 62: 2168–2173, 1987.
29. Duchateau, J. and Hainaut, K.Effects of immobilization on contractile properties recruitment and firing rates of human motor units. *Journal of Physiology* 422: 55–65, 1990.
30. Duchateau, J. and Hainaut, K.Effects of immobilization on electromyogram power spectrum changes during fatigue. *European Journal of Applied Physiology and Occupational Physiology* 63: 458–462, 1991.
31. Dudley, G.A., Duvoisin, M.R., Adams, G.R., Meyer, R.A., Belew, A.H., and Buchanan, P.Adaptations to unilateral lower limb suspensions in humans. *Aviation, Space, and Environmental Medicine* 63: 678–683, 1992.
32. Eguchi, Y., Jinde, M., Murooka, K., Konno, Y., Ohta, M., and Yamato, H. Stretching versus transitory icing: which is the more effective treatment for attenuating muscle fatigue after repeated manual labor? *Eur J Appl Physiol* 114: 2617–2623, 2014.
33. Esola, M.A., McClure, P.W., Fitzgerald, G.K., and Siegler, S.Analysis of lumbar spine and hip motion during forward bending in subjects with and without a history of low back pain. *Spine (Phila Pa 1976)* 21: 71–78, 1996.
34. Farfan, H.F.The biomechanical advantage of lordosis and hip extension for upright activity. Man as compared with other anthropoids. *Spine (Phila Pa 1976)* 3: 336–342, 1978.

35. Folha, P.N., Bacurau, R.P.F., Aoki, M.S., and Massa, M.Effect of active static stretching on glycemia in type 2 diabetes. *Manual Therapy, Posturology and Rehabilitation Journal* 22: 1–16, 2024.
36. Fradkin, A.J., Gabbe, B.J., and Cameron, P.A.Does warming up prevent injury in sport? The evidence from randomised controlled trials? *J Sci Med Sport* 9: 214–220, 2006.
37. Fuglsang-Fredriksen, A. and Scheel, U.Transient decrease in number of motor units after immobilisation in man. *Journal of Neurology, Neurosurgery, and Psychiatry* 41: 924–929, 1978.
38. Godges, J.J., MacRae, H., Longdon, C., and Tinberg, C.The effects of two stretching procedures on the economy of walking and jogging. *Journal of Orthopaedics and Sport Physical Therapy* 7: 350–357, 1989.
39. Goldspink, G., Tabary, C., Tabary, J.C., Tardieu, C., and Tardieu, G.Effect of denervation on the adaptation of sarcomere number and muscle extensibility to the functional length of the muscle. *J Physiol* 236: 733–742, 1974.
40. Goldspink, G. and Williams, P.E.The nature of the increased passive resistance in muscle following immobilization of the mouse soleus muscle [proceedings]. *J Physiol* 289: 55P, 1979.
41. Golub, L.J.Exercises that alleviate primary dysmenorrhea. *Contemporary Ob/Gyn* 5: 51–59, 1987.
42. Golub, L.J., Menduke, H., and Lang, W.R.Exercise and dysmenorrhea in young teenagers: A 3-year study. *Obstet Gynecol* 32: 508–511, 1968.
43. Golub, L.J., Lang, W.R., and Menduke, H.Dysmenorrhea in high school and college girls: Relationship to sports participation. *Western Journal of Surgery, Obstetrics and Gynecology* 3: 163–165, 1958.
44. Halka, A.T., Turner, N.J., Carter, A., Ghosh, J., Murphy, M.O., Kirton, J.P., Kielty, C.M., and Walker, M.G.The effects of stretch on vascular smooth muscle cell phenotype in vitro. *Cardiovasc Pathol* 17: 98–102, 2008.
45. Halkjaer-Kristensen, J. and Ingemann-Hansen, T.Wasting of the human quadriceps muscle after knee ligament injuries. IV Dynamic and static muscle function. *Scand J Rehab Med Suppl* 13: 29–37, 1985.
46. Halkjaer-Kristensen, J. and Ingemann-Hansen, T.Wasting of the human quadriceps muscle after ligament injuries. I Anthropometrical consequences. *Scand J Rehab Med Suppl* 13: 5–10, 1985.
47. Hellebrandt, F.A.Cross education: Ipsilateral and contralateral effects of unimanual training. *JAP* 4: 136–144, 1951.
48. Herda, T.J., Cramer, J.T., Ryan, E.D., McHugh, M.P., and Stout, J.R.Acute effects of static versus dynamic stretching on isometric peak torque, electromyography, and mechanomyography of the biceps femoris muscle. *J Strength Cond Res* 22: 809–817, 2008.
49. Hortobagyi, T.Cross Education and the Human Central Nervous System: Mechanisms of Unilateral Interventions Producing Contralateral Adaptations. *IEEE Engineering in Medicine and Biology* January/February: 22–28, 2005.
50. Howes, R.G. and Isdale, I.C.The non-myotendious force transmission. *Rheumatology and Physical Medicine* 2: 72–77, 1971.
51. Inami, T., Baba, R., Nakagaki, A., and Shimizu, T.Acute changes in peripheral vascular tonus and systemic circulation during static stretching. *Res Sports Med* 23: 167–178, 2015.
52. Jamtvedt, G., Herbert, R.D., Flottorp, S., Odgaard-Jensen, J., Havelsrud, K., Barratt, A., Mathieu, E., Burls, A., and Oxman, A.D.A pragmatic randomised trial

of stretching before and after physical activity to prevent injury and soreness. *Br J Sports Med* 44: 1002–1009, 2010.

53. Kato, M., Nihei Green, F., Hotta, K., Tsukamoto, T., Kurita, Y., Kubo, A., and Takagi, H.The Efficacy of Stretching Exercises on Arterial Stiffness in Middle-Aged and Older Adults: A Meta-Analysis of Randomized and Non-Randomized Controlled Trials. *Int J Environ Res Public Health* 17, 2020.
54. Kibele, A., Granacher, U., Muehlbauer, T., and Behm, D.G.Stable, Unstable and Metastable States of Equilibrium: Definitions and Applications to Human Movement. *J Sports Sci Med* 14: 885–887, 2015.
55. Konrad, A., Alizadeh, S., Daneshjoo, A., Anvar, S.H., Graham, A., Zahiri, A., Goudini, R., Edwards, C., Scharf, C., and Behm, D.G.Chronic effects of stretching on range of motion with consideration of potential moderating variables: A systematic review with meta-analysis. *J Sport Health Sci*, 2023.
56. Konrad, A., Nakamura, M., Warneke, K., Donti, O., and Gabriel, A.The contralateral effects of foam rolling on range of motion and muscle performance. *Eur J Appl Physiol* 123: 1167–1178, 2023.
57. Kruse, N.T. and Scheuermann, B.W.Effect of self-administered stretching on NIRS-measured oxygenation dynamics. *Clin Physiol Funct Imaging* 36: 126–133, 2016.
58. Kubo, K., Kanehisa, H., and Fukunaga, T.Effects of transient muscle contractions and stretching on the tendon structures in vivo. *Acta Physiol Scand* 175: 157–164, 2002.
59. Lima, T.P., Farinatti, P.T., Rubini, E.C., Silva, E.B., and Monteiro, W.D.Hemodynamic responses during and after multiple sets of stretching exercises performed with and without the Valsalva maneuver. *Clinics (Sao Paulo)* 70: 333–338, 2015.
60. Locke, J.C.Stretching away from back pain, injury. *Occup Health Saf* 52: 8–13, 1983.
61. Lundberg, G. and Gerdle, B.Correlations between joint and spinal mobility, spinal sagittal configuration, segmental mobility, segmental pain, symptoms and disabilities in female homecare personnel. *Scand J Rehabil Med* 32: 124–133, 2000.
62. Magnusson, S.P., Simonsen, E.B., Aagaard, P., Sorensen, H., and Kjaer, M.A mechanism for altered flexibility in human skeletal muscle. *Journal of Physiology* 497 (Pt 1): 291–298, 1996.
63. McDonough, A.L.Effects of immobilization and exercise on articular cartilage – A review of the literature. *The Journal of Orthopaedic and Sports Physical Therapy* 1: 2–5, 1981.
64. McGill, S.M.Stability: From biomechanical concepts to chiropractic practice. *Journal of the Candain Chiropractic Association* 2: 75–88, 1999.
65. McHugh, M. and Hogan, D.Effect of knee flexion angle on active joint stiffness. *Acta Physiologica Scandinavica* 180: 249–254, 2004.
66. McHugh, M.P. and Cosgrave, C.H.To stretch or not to stretch: the role of stretching in injury prevention and performance. *Scand J Med Sci Sports* 20: 169–181, 2010.
67. McHugh, M.P. and Nesse, M.Effect of stretching on strength loss and pain after eccentric exercise. *Med Sci Sports Exerc* 40: 566–573, 2008.
68. Miller, K.C., Harsen, J.D., and Long, B.C.Prophylactic stretching does not reduce cramp susceptibility. *Muscle Nerve*, 2017.

69. Pope, R., Herbert, R., and Kirwan, J.Effects of ankle dorsiflexion range and pre-exercise calf muscle stretching on injury risk in army recruits. *Australian Physiotherapy* 44: 165–177, 1998.
70. Pope, R., Herbert, R., Kirwan, J., and Graham, B.A randomized trial of pre-exercise stretching for prevention of lower-limb injury. *Medicine and Science in Sports and Exercise* 32: 271, 2000.
71. Rasch, P.J. and Burke, J. *Kinesiology and applied anatomy.* Philadelphia, PA: Lea & Febiger, 1989.
72. Russell, G.S. and Highland, T.R. *Care of the lower back.* Columbia, MO: Spine, 1990.
73. Sahrmann, S.A. *Diagnosis and treatment of movement impairment syndromes.* St. Louis, MS: Mosby Publishers Inc., 2002.
74. Sale, D.G., McComas, A.J., MacDougall, J.D., and Upton, A.R.M.Neuromuscular adaptation in human thenar muscles following strength training and immobilization. *Journal of applied physiology* 53: 419–424, 1982.
75. Sanfilippo, D., Beaudart, C., Gaillard A., Bornheim, S., Bruyere, O., and Kaux, J. F.What Are the Main Risk Factors for Lower Extremity Running-Related Injuries? A Retrospective Survey Based on 3669 Respondents. *Orthop J Sports Med* 9: 23259671211043444, 2021.
76. Shima, N., Ishida, K., Katayama, K., Morotome, Y., Sato, Y., and Miyamura, M. Cross education of muscular strength during unilateral resistance training and detraining. *European Journal of Applied Physiology* 86: 287–294, 2002.
77. Shrier, I.Does stretching improve performance?: a systematic and critical review of the literature. *ClinJSport Med* 14: 267–273, 2004.
78. Small, K., McNaughton, L., and Matthews, M.A systematic review into the efficacy of static stretching as part of a warm-up for the prevention of exercise-related injury. *Res Sports Med* 16: 213–231, 2008.
79. Tabary, J.C., Tabary, C., Tardieu, C., Tardieu, G., and Goldspink, G.Physiological and structural changes in the cat's soleus muscle due to immobilization at different lengths by plaster casts. *J Physiol* 224: 231–244, 1972.
80. Tabary, J.C., Tardieu, C., Tabary, C., Lombard, M., Gagnard, L., and Tardieu, G. Readaptation of the number of sarcomeres in series in isolated soleus fibers after neurectomy. *J Physiol (Paris)* 65: Suppl 3:509A, 1972.
81. Thomas, E., Bellafiore, M., Gentile, A., Paoli, A., Palma, A., and Bianco, A.Cardiovascular Responses to Muscle Stretching: A Systematic Review and Meta-analysis. *Int J Sports Med*, 2021.
82. Thorpe, C.T., Godinho, M.S.C., Riley, G.P., Birch, H.L., Clegg, P.D., and Screen, H.R.C.The interfascicular matrix enables fascicle sliding and recovery in tendon, and behaves more elastically in energy storing tendons. *J Mech Behav Biomed Mater* 52: 85–94, 2015.
83. Thorpe, C.T., Udeze, C.P., Birch, H.L., Clegg, P.D., and Screen, H.R.Capacity for sliding between tendon fascicles decreases with ageing in injury prone equine tendons: a possible mechanism for age-related tendinopathy? *Eur Cell Mater* 25: 48–60, 2013.
84. Toft, E., Espersen, G.T., Külund, S., Sinkjër, T., and Hornemann, B.C.Passive tension of the ankle before and after stretching. *The American Journal of Sports Medicine* 17: 489–494, 1989.
85. Waddell, G., Somerville, D., Henderson, I., and Newton, M.Objective clinical evaluation of physical impairment in chronic low back pain. *Spine (Phila Pa 1976)* 17: 617–628, 1992.

86. Webber, S. and Kriellaars, D.Neuromuscular factors contributing to in vivo eccentric moment generation. *J Appl Physiol (1985)* 83: 40–45, 1997.
87. Weldon, S.M. and Hill, R.H.The efficacy of stretching for prevention of exercise-related injury: a systematic review of the literature. *Man Ther* 8: 141–150, 2003.
88. White, M.J. and Davies, C.T.M.The effects of immobilization, after lower leg fracture, on the contractile properties of human triceps surae. *Clinical Science* 66: 277–282, 1984.
89. White, M.J., Davies, C.T.M., and Brooksby, P.The effects of short-term voluntary immobilization on the contractile properties of the human triceps surae. *J Gen Physiol* 69: 21–27, 1984.
90. Witvrouw, E., Mahieu, N., Danneels, L., and McNair, P.Stretching and injury prevention: an obscure relationship. *Sports Med* 34: 443–449, 2004.
91. Witvrouw, E., Mahieu, N., Roosen, P., and McNair, P.The role of stretching in tendon injuries. *Br J Sports Med* 41: 224–226, 2007.
92. Wong, A. and Figueroa, A.Effects of Acute Stretching Exercise and Training on Heart Rate Variability: A Review. *J Strength Cond Res* 35: 1459–1466, 2021.
93. Woods, K., Bishop, P. and Jones, E.Warm-up and stretching in the prevention of muscular injury. *Sports Med* 37: 1089–1099, 2007.
94. Yamamoto, K., Kawano, H., Gando, Y., Iemitsu, M., Murakami, H., Sanada, K., Tanimoto, M., Ohmori, Y., Higuchi, M., Tabata, I., and Miyachi, M.Poor trunk flexibility is associated with arterial stiffening. *Am J Physiol Heart Circ Physiol* 297: H1314−H1318, 2009.
95. Zakaria, A.A., Kiningham, R.B., and Sen, A.Effects of Static and Dynamic Stretching on Injury Prevention in High School Soccer Athletes: A Randomized Trial. *J Sport Rehabil* 24: 229–235, 2015.
96. Zhang, W. and Bai, N.The role of functional dynamic stretching training in dance sports. *Review Brazil Medicine Esporte* 28: 837–839, 2022.
97. Zhou, S.Chronic neural adaptations to unilateral exercise: mechanisms of cross education. *Exerc Sport Sci Rev* 28: 177–184, 2000.

9 Does Stretching Affect Performance?

Static and PNF Stretching

At least from the mid-20th century, stretching was believed to enhance performance. An early study by the renowned George Dintiman reported that stretch training enhanced sprint speed (45). Higher baseline levels of flexibility in volleyball players were associated with better performance in agility, acceleration speed, and lower body muscular power as compared to players with low flexibility level (67). But a variety of studies have indicated no beneficial effects of stretch training on the oxygen cost of running (three days/week for ten weeks) (110), drop jumps (four days/week for six weeks) (153), vertical jumps (four days/week for six weeks) (8), or sprint speed (8). However static stretching two days/week for 8 weeks did improve rebound bench press performance (145). But the acute effect of static stretching within a pre-event warm-up has received far more attention lately.

It was in the 1990s that more frequent investigations began into the role of stretching as part of a warm-up prior to training or competition. The common thought since the 1960s was that static stretching would improve performance and Worrell and colleagues in 1994 (147) did report that four hamstring stretches of 15–20 seconds each improved eccentric and concentric leg flexion torque, as expected! As mentioned previously, the acute increase in range of motion (ROM) was thought to reduce resistance to movement, and provide greater limb excursions that would be important for work output (work = force x linear or angular distance), running stride length, activities that need substantial flexibility (e.g., gymnastics, dance, figure skating, and others) and other movements. The proposition that improved flexibility would improve performance arises from the concepts of stress, strain, stiffness and the modulus of elasticity. Stress and strain are related as they are defined either as the applied force divided by the tissue cross-sectional area (stress) or the ratio or change in tissue length with an applied force (strain). Stress and strain are directly related to tissue stiffness, as stiffness is defined as the ratio of stress to strain or the ratio of changes in force to changes in length. Often the term compliance is used as the opposite to stiffness. A more complaint muscle or tendon is a less stiff muscle or tendon. Historically, Robert Hooke

DOI: 10.4324/9781032709086-10

found that there was a proportional relationship between force and elongation. According to his modus of elasticity, stiffer tissues need a greater stress (force or load) to produce a greater strain (change in length). Thus, a more compliant or less stiff musculotendinous unit should need less force or load to move it through a ROM resulting in a greater economy of movement and less resistance to movement.

But Kokkonen and colleagues in 1998 were among the first to contradict the prevailing notion that increased flexibility would improve performance. They found that five stretches involving six repetitions of 15 seconds each decreased knee flexion and extension force by 7–8 percent (92). A 1999 study by Avela et al. (3) had subjects receive one hour of repeated passive static stretching reporting impairments in force, electromyography (EMG) and reflex sensitivity. Typical of the problem with many of these earlier studies is the real-world relevance. What percentage of the population stretches their calf muscles for 60 minutes?

This led to many more studies in the early part of the millennium reporting stretching-induced impairments. A number of comprehensive reviews have reported that prolonged static stretching can lead to subsequent impairments in strength, power, sprint speed, agility, balance, evoked contractile properties and other performance and physiological parameters (12, 15, 87, 127). These impairments on average range from 4–7.5 percent. Static stretching can even impair the extent of muscle hypertrophy, strength and insulin-like growth factor-1 (IGF-1) (23) achieved if you consistently stretch before or during resistance training. These reports have led to changes in how athletes prepare or warm-up for training and competition. Whereas static stretching was widespread prior to the new millennium, there has been a shift away from static stretching to more dynamic stretching and dynamic activities. A review of American college tennis coaches in a 2012 article showed that 87 percent of coaches always employed a warm-up and within that warm-up, 38.5 percent used a combination of static and dynamic stretching, 10.5 percent only used static stretching and the rest relied primarily on dynamic activities and dynamic stretching (84). The results were slightly different for American university cross country and track and field coaches in a 2013 article with 98.1 percent always ensuring a warm-up and within that group 44.7 percent used a combination of static and dynamic stretching whereas 41.5 percent only performed dynamic stretching and 10.5 percent used static or proprioceptive neuromuscular facilitation (PNF) stretching (86). Their throwing counterparts used a warm-up 95.6 percent of the time with 40.7 percent using dynamic stretching, 38.5 percent employing a combination of static and dynamic stretching and 11.1 percent using static or PNF stretching (85). A French survey of 3,546 individuals (47.3 percent women and 52.7 percent men, national/international level athletes: 25.2 percent, regional level athletes: 29.8 percent, or recreationally active: 44.9 percent) reported that stretching was generally performed either after training (72.4 percent) or as a warm-up (49.9 percent), with static stretching primarily utilized (88.2 percent) except for the warm-up when most people used dynamic stretching (86.2 percent) (4). So, while the warm-up paradigm has shifted to an emphasis on dynamic stretching or a combination, about 10 percent of coaches in these studies still use static stretching alone. Is this paradigm shift a legitimate movement based on irrefutable fact or one of the many fads that we see come and go over the decades?

With static stretching-induced impairments, the emphasis should be placed on the term "prolonged" static stretching. It is generally agreed that 60 seconds or more of static stretching per muscle group is more likely to cause significant performance decrements (12, 15, 87). Muscle length is also a mitigating factor. Static stretches performed at short muscle lengths showed substantially greater deficits (10.2 percent), while static stretching at long muscle lengths actually resulted in small strength gains (2.2 percent) (12).

PNF stretching results from our review (11 PNF studies) reported approximately 4 percent performance deficits (12). An earlier review (2007) found no clear evidence for a negative effect of PNF stretching on subsequent exercise performance (121). Sarah Marek and colleagues found that both static and PNF stretching led to similar strength, power, and muscle activation impairments but the magnitude of the deficits were small (104). Therefore, as usual, there is never complete agreement in science. There are actually some studies that show PNF-induced improvements in isokinetic torque (147) and postural stability (123). But contradicting those studies are others reporting either no changes in strength or jump height (151) or PNF-induced deficits in proprioception (132), movement time, hip flexion angular velocity (101), vertical jump height (24, 33), strength, power and muscle activation (104). One study showed that ten repetitions of PNF stretching increased blood lactate concentration indicating there would be a greater reliance on anaerobic metabolism following PNF stretching (69). In the few studies that compared static and PNF stretching, PNF presented greater (6.4 percent vs. 2.3 percent) subsequent deficits than the static stretching protocols. An important note to keep in mind is that most of these studies intervened only with prolonged stretching without the additional activity you typically experience in a warm-up. The methodological problems and misinterpretations of these types of studies will be examined later. But if you only static or PNF stretch and do not have time to commit to a full warm-up procedure, how important are the consequences?

While a 5 percent deficit (15) might not seem that dramatic to the average athlete, it can be devastating to an elite athlete. Usain Bolt won the 100-metre gold medal at the 2012 Olympic Games in 9.63 seconds. If a competitor was 5 percent slower, then where would he place in the competition. In fact, Gerard Phiri from Zambia was almost exactly 5 percent slower and ran the 100 metres in 10.11 seconds. Did Phiri medal? Did Phiri make it to the finals? No, Phiri came in second last place (15th) in the semi-finals. The 5 percent difference was devastating. Bolt makes millions of dollars from competing and endorsements. and I can bet that 99.9 percent of the readers have never heard of Phiri. Adidas, Nike, or Reebok are not knocking at his door throwing money at him. Perhaps you watched the 2016 Olympics and saw Bolt win again in 9.81 seconds (he was getting older and slowed down!). As a Canadian, I was cheering for Andre De Grasse. He won a bronze medal in 9.91 seconds (see Figure 9.1). The difference between Bolt and De Grasse was 1.009 percent. Performing the correct warm-up, is unbelievably crucial in elite competition. In our 2011 review (15) we stated, "It would be wise to be cautious when implementing static stretching of any duration or for any population when high-speed, rapid stretch-shortening cycle, explosive or reactive forces are necessary, particularly if any decreases in performance, however small, would be important." However, there are a number of caveats to this statement. We also mentioned that "All individuals should include static stretching in their overall fitness and wellness activities for the health and functional benefits associated with increased ROM and musculotendinous compliance." In our 2016 review (12), which involved a more expansive and critical examination of the literature, we modified our views and said the following, "When a typical pre-event warm-up is complete (i.e., initial aerobic activity, stretching component, 5–15 minutes of task- or activity-specific dynamic activities), the benefits of SS and PNF stretching for augmented ROM and reduced injury risk balance, or may outweigh, any possible cost of (trivial) performance decrements." Sounds very confusing! Should we stretch or not before explosive or high-speed training or competition?

To simplify the answer, it might be easier to think of yourself either as a Ferrari or a Cadillac (see Figure 9.2). Elite athletes are the Ferrari, Lamborghini or Bugati sport cars of the world. The purpose of the elite athlete is to push themselves to their physiological, anatomical and psychological limits.

Figure 9.1 Olympic sprint final 2016

Figure 9.2 Ferrari (tighter suspension) versus Cadillac (more compliant, less stiff suspension)

They need to produce maximal forces, or power, and many times in the briefest possible periods exerting extremely high pressures, torques, impacts and other stresses to their body. Usain Bolt and the other international-calibre sprinters have foot contact times on the track of less than 100 milliseconds. Thus, they must hit the ground and take-off without absorbing and losing much energy. Elite sprinters will leave the starting blocks in about 120–160 milliseconds (107). The fastest start was by Bruny Surin (another Canadian!) at 101 milliseconds in the 1999 world championship semi-final. Athletes with stiffer or less compliant musculotendinous systems are reported to be more economical runners (less energy expended for the same amount of work) than their more flexible counterparts (59, 83, 140). A number of studies have reported that static stretching can reduce sprinting speed (9, 31, 55, 125, 128, 146). Similar to human sprinters, a Ferrari race car needs to accelerate from the start and move out of curves and corners rapidly. To accomplish this task, race cars have tight suspension so they react to the road immediately. Driving a Ferrari over a bumpy road would not be much fun and in fact excruciatingly painful for your lower back as the suspension would not absorb much of the shock of the pot holes in the roads. For the average city or country road you want a car with nice soft suspension that absorbs the shocks so that the driver feels almost nothing when encountering a bump in the road. A Cadillac and other similar cars have great suspensions that effectively absorb these shocks. You would never expect a Cadillac to beat a Ferrari in a race especially on a route that included curves and turns. After every race, mechanics must examine, repair or replace the damaged components of race cars. There are always damaged parts after every race. The car has been pushed to its mechanical limit and the suspension did not adequately absorb or reduce the effect of the road environment on the other parts of the Ferrari. You would understandably be very upset, if every time you drove your Cadillac you had to bring it to the mechanic to get it repaired. The top sprinters (Bolt and colleagues), tennis players (Roger Federer, Novak Djokovic, and colleagues) and other individual sport athletes as well as team sport athletes have athletic trainers, physiotherapists, massage therapists and others ready to provide them with treatment immediately after a competition when they have pushed their bodies to the limit. The average fitness enthusiast and weekend warrior athlete does not have a cadre of health practitioners to treat him/her after every workout. The majority of the population does not push themselves to the extreme limit.

There is no single correct prescription for the optimal degree of flexibility. Everybody has their "Goldilocks" zone. The elite sprinter needs to transfer contractile forces as quickly as possible from the muscle to the tendon to the bone resulting in movement. If a tendon is overly slack (high compliance) there will be a negative impact on electromechanical delay (EMD). The EMD is the time it takes for the muscle to be activated, contract, pull and tighten the tendon and then have the bone move. Under laboratory conditions the EMD is measured as the duration between the onset of the muscle action potential wave (M-wave) to the onset of force or torque production (see Figure 9.3). If an elite athlete's parallel (muscle membranes) and series elastic (i.e., tendons) components were very compliant then the EMD is prolonged and the time it takes for a muscle contraction to actually result in movement is also lengthened. This of course is anathema to the elite sprinter who wants to explode out of the blocks in the shortest time (reaction time) or spend the least amount of time in contact with the track (increased stride rate).

Previously, I told you that prolonged static stretching or increased flexibility can decrease running economy. But there are also studies that contradict these previous studies with reports that static stretching also decreased the energy cost of running (60) or had no effect (72). How is this possible? Were one set of scientists incompetent and incorrect? No, it is necessary to look at the specificity of

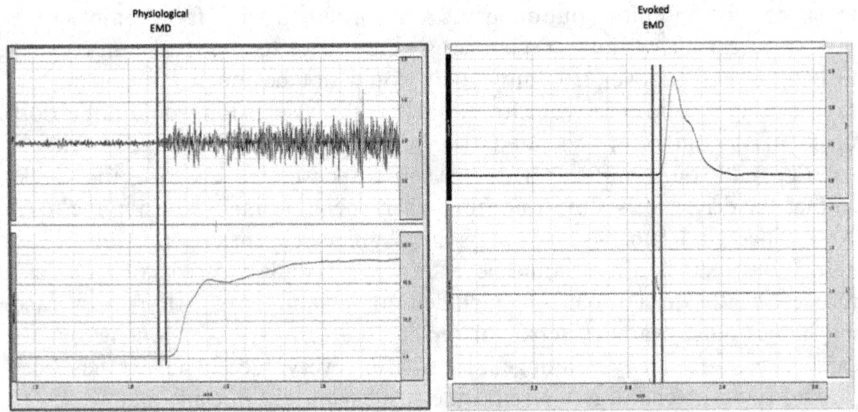

Figure 9.3 Electromechanical delay (EMD). Physiological EMD is the time from the onset of muscle activation (illustrated with electromyography in the first figure) and the onset of voluntary force. Evoked EMD is the time from the onset of evoked activation (illustrated by the electrically evoked muscle action potential in the second figure) to the onset of evoked twitch force. EMD in these examples are approximately 40–60 milliseconds

the task. Schmidtbleicher (126) defined two types of rebound jumps or stretch shortening cycle (SSC) actions (see Figures 9.4 and 9.5). A jump performed with a contact time of less than 250 milliseconds was classified as a short SSC and more than 250 milliseconds was considered a long SSC. The two studies that found improved running economy with prior stretching used submaximal, slower, prolonged running (i.e., 40, 60, 80 percent of VO2 maximum or below lactate threshold) with long SSC. With slower running and longer SSC, the person is in longer contact with the ground (see Figure 9.6). A very stiff musculotendinous system absorbs the reaction forces over a very short duration and returns the energy to the muscles very quickly (desirable for sprinters). If the system is very stiff, the mechanical and reflex energy return (stretch-shortening cycle) would occur while the slower runner is still in the landing phase and not ready to propel themselves into the next stride. A more compliant or flexible muscle and tendon would absorb the reaction forces over a longer period and the spring-like action of the muscles and tendons would return the energy to the muscles when the distance runner is ready to push off the ground.

Studies that examine evoked contractile properties (electrical stimulation of the muscle or nerve) might give some insight into some of the mechanisms underlying static stretch-induced impairments. A few studies have reported static stretch-induced decreases in twitch force, rate of force development and prolonged EMD (29, 36, 52, 122). As a reminder, EMD is primarily affected by the musculotendinous unit (MTU) compliance or stiffness. It is suggested that there is a direct relationship between EMD and muscle stiffness (63). If the MTU is more compliant it will take longer to transfer the energy of the myofilament cross-bridge kinetics to the muscle and tendon tissues to move the bone. Muscle stiffness has a greater influence on athletic performance

Figure 9.4 Stretch-shortening cycle: Hurdle jumps (from right to left) beginning with take-off (propulsion), flight phase, (clearance of the hurdle), landing (involving eccentric contractions to absorb the reaction forces), with a short contact (amortization or transition phase) time, followed by concentric contractions for another propulsion phase. The eccentric contractions during landing stiffen the musculotendinous unit (MTU) to decrease electromechanical delay while elongation of the MTU activates muscle spindle stretch receptors (nuclear bag and chain fibres) to induce reflex contractions to augment the force of the subsequent concentric contractions

Figure 9.5 Stretch-shortening cycle: Drop jump preparation, landing (amortization or transition phase with elongation/stretching of muscles and tendons) and take-off (propulsion: concentric contractions)

than flexibility (120). The intrinsic or baseline stiffness of a muscle can still be moderately high and yet with a high tolerance to stretch (pain or discomfort), an individual can still have substantial levels of flexibility.

Some researchers have compared a mechanomyogram signal to the force or EMG signal to parcel out the EMD associated with excitation contraction coupling and MTU viscoelastic properties. A mechanomyogram (MMG) monitors changes in the muscle shape by sensing vibrations and oscillations of the muscle fibres at the resonant frequency of the muscle. With a higher

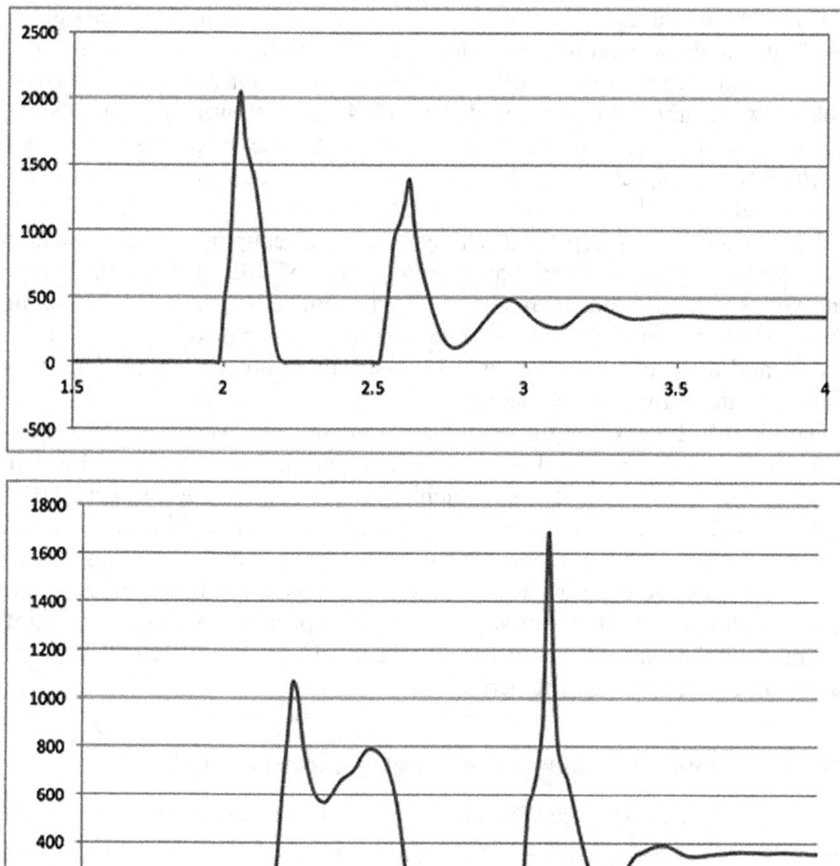

Figure 9.6 Drop jump stretch-shortening cycle reaction forces with a short versus long contact time. The first peak is the landing from a drop jump (contact time), followed by the flight phase (second baseline) and the last peak is the landing following the flight phase

signal-to-noise ratio than the surface EMG it is purported to be able to monitor muscle activity from deeper muscles. After five sets of 45 seconds of passive stretching, the excitation-contraction (E-C) coupling aspect of the EMD (Δt: EMG-MMG) was prolonged for approximately 15 minutes, whereas the MTU visco-elasticity (Δt: MMG-tetanic force) was protracted for up to two hours (51). It is necessary to remind the reader that research is almost never unanimous. A study from our lab also found force deficits up to two hours after stretching two muscles for three sets of 45 seconds each. However, while there were significant voluntary force deficits (10 percent), the evoked twitch and tetanic force deficits (1–4 percent) were non-significant over that same period (114). However, peak tetanic force has also been reported to be depressed for two hours after five repetitions of 45-second static stretches with 15 seconds' rest between stretches (52). Changes in evoked twitch forces represent changes in E-C coupling (muscle action potential leading to the release and sequestration of calcium from the sarcoplasmic reticulum) and tetanic forces represent changes in myofilament kinetics. Hence, five x 45 seconds of stretching in this one study damaged the myosin and actin cross bridges, which would affect not only peak force but likely the rate of force development.

The Goldilocks flexibility or ROM zone for sprinters versus distance runners would be different based on their differing ground contact times. Further evidence for task specificity of stretching is provided by experiments that demonstrated a 5 percent increase in rebound chest press (145) and no significant effects on isometric bench press (136). As the amortization or transition phase from the eccentric to the concentric phases of a bench press would be much slower than the contact time during sprinting, a more compliant system would be able to absorb and return the elastic energy in a more appropriate task specific time period.

Effects of Static Stretching on the Stretch-Shortening Cycle

The stretch-shortening cycle (SSC) is an important neuromuscular mechanism that enhances power production with rapid, explosive activities and contractions. Prime examples of SSC activities are plyometric activities such as countermovement, hurdle and drop jumps, hopping, bounding, skipping, sprinting, and other similar activities. With SSC activities, external forces (e. g., body mass accelerated by gravity) lengthen the muscle during an active contraction (eccentric phase), which is followed by a concentric (shortening) contraction (93, 112). SSC exercises use the elastic energy that was stored during the eccentric contraction phase to augment the concentric phase (28, 94). This augmentation involves both mechanical (musculotendinous) and neural (reflexive) components. Like an elastic band, the contractile elements (i.e., myosin) and muscle connective tissue (i.e., titin, desmin) and tendons are stretched resulting in the storage of potential elastic energy that is returned to the neuromuscular system shortly thereafter. The energy stored in the system can be extremely high. For example, a study from our laboratory reported that hurdle jumps that emphasized short contact times generated mean reaction forces of 4,335 Newtons (442 kg or 972 lbs). This reaction force was approximately 5.5 times the body mass and approximately 100 percent higher than the much slower SSC action of countermovement jumps (27). The elongation of the MTU does not need to be extensive to achieve this elastic rebound. A tendon elongation of 6–7 percent during the stretching phase is sufficient to elicit the SSC. Rack and Westbury (118) originally labelled this small elongation: short-range elastic stiffness. If the tendon can return elastic energy with only a 6–7 percent elongation, we can see why sprinters may not need or want a very compliant MTU. A stiffer MTU is still fine for transferring elastic recoil energy.

The duration for the return of this elastic recoil is dependent upon the compliance or elasticity of the SSC. A tighter system will return this elastic recoil earlier than a more compliant system. Slow SSC actions only get a minor contribution from the passive elastic elements, creating a greater reliance on muscle contractile properties to produce concentric phase force. However, a rapid SSC can take greater advantage of elastic energy from the MTU (39, 80). With rapid SSC actions, the cross bridges can remain attached during the stretch, to augment the force during the concentric phase (6). The rapid deceleration during the eccentric stretch phase followed by the acceleration during the propulsive, concentric, phase contributes to a power potentiation and improved energy economy that is dependent on the speed of the eccentric stretch phase (80).

Not only is there a mechanical elastic storage and energy return but the SSC causes an excitation of the muscle spindle reflexes (111). Rapid SSC actions induce greater reaction forces, muscle reflex activity and muscle activation (137). The muscle spindles sense and react to changes in the extent and rate of muscle lengthening. Within the muscle spindle, nuclear chain fibres preferentially respond to changes in the amount of stretch or elongation of the muscle while the nuclear bag fibres respond to both the extent of elongation and the rate. These signals return to the spinal motoneurons via annulospiral (high conduction velocity) and flower spray endings (low conduction velocity) through Ia (nuclear bag and chains) and II (nuclear chains) afferents respectively. A monosynaptic myotactic reflex is initiated from the rapid stretch action resulting in the depolarization/activation (contraction) of the alpha motoneuron of the stretched muscle and inhibition (di-synaptic) of the antagonist muscle (105). This reflex action aids in neural potentiation, stiffening of the contractile components due to the myofilament crossbridge attachments and augmenting the active contraction forces.

The spindle afferents can contribute up to 30 percent to the activation of the motoneurons (57). Persistent inward currents (PIC) from prior contractions contribute to motoneuron depolarization during the subsequent contractions. The PIC is a depolarizing inward current of motoneurons that respond to brief synaptic input with prolonged firing activity, even after the input stops. (known as self-sustained firing). They can also increase the maximum motoneuron firing rate. PICs are derived from Na+ and low voltage-activated L-type Ca2+ ion channels. A small afferent signal that activates the dendritic PIC is amplified creating a larger current to the soma (motoneuronal body) that produce higher motoneuronal firing rates

(50). It is interesting that neurophysiologists use the term "warm-up" for the increased activation of the Ca^{++} PIC. Thus, the increased reflex afferent activity from repeated SSC actions in concert with supraspinal drive would potentiate the motoneurons response (i.e., greater rate coding). However, a number of static stretching studies have reported H-reflex (afferent excitability of the spinal motoneuron) inhibition with stretching (3, 53, 64, 65, 66, 133). In contrast, work from the Blazevich lab by Pulverenti et al. reported a lack of motor evoked potential (MEP) (115, 117) or H-reflex amplitudes (117) changes following stretching in the soleus (five x 60 seconds constant torque static stretches) during submaximal contractions, suggesting that changes in the excitability of the corticospinal-motoneuronal (115, 117) and Ia afferent spinal reflex (117) pathways do not contribute to post-stretch neural impairments. So, is there cortical, spinal, PIC, or afferent excitability inhibition or not with stretching? Two years after the Pulverenti publication, the Trajano and Blazevich team published a narrative review on the topic (138). They divulged that there is clear evidence for prolonged static stretch-induced neural drive reduction and their conclusion was that the reduction in motoneuron excitability caused by a reduction in PIC strength was the most likely causative factor. PIC reduction was dependent on reduced facilitation by monoaminergic neuromodulatory systems (i.e., noradrenaline). Hence, static stretching moves the sympathetic-parasympathetic balance towards a more parasympathetic relaxation, negatively influencing motoneurons' ability to maximally fire resulting in lower maximal muscle force (138). These conclusions are in accord with a prior narrative review by the Trajano, Nosaka, and Blazevich crew (139) when they highlighted that primary factors inducing prolonged static stretch-induced force impairments are related to reductions in the central nervous system motor command as well as a decreased amplification of the spinal motor command.

There are many explosive, power type athletes such as gymnasts and figure skaters who have extensive flexibility and yet still have a rapid and efficient SSC. Does this not contradict this previous discussion? These types of athletes can perform powerfully even with a very compliant MTU if they actively stiffen the system before contracting. Hence it is important for them to pre-activate the system. More experienced and accomplished athletes will pre-activate the muscles with a feedforward (pre-planned) central command (44). The first 40 milliseconds of ground contact time is too soon for spindle reflexes to stiffen the system and thus pre-activation from supraspinal activation is necessary (61).

Stretching Effects on Balance

While we often connect the importance of improved balance to athletic performance, we have to remember that slips, trips, and falls contribute to 95 percent of hip fractures in seniors (73) and are the most common cause of traumatic brain injury (81). There is some evidence that improved dorsiflexion ROM can facilitate balance test performance (anterior direction in the Star Excursion Balance Test: SEBT) (77) and even with 10–30 minutes of static or dynamic stretching (15-second repetitions) with a five-minute aerobic warm-up there was a small magnitude improvement in SEBT performance (19). The SEBT is usually performed under controlled and slower conditions and thus a less stiff, more compliant system that reacts over a more prolonged period could provide better feedback and postural adjustments for this test. The improved SEBT performance with greater flexibility might not necessarily translate to the far more rapid adjustments to posture (metastability) encountered with such high velocity perturbations found with downhill skiing, rugby, American football, and other sports or everyday activities like slipping on an icy sidewalk. Our narrative review (18) reported on balance impairments with static stretching durations of 30 seconds up to five minutes (six studies) contrasting with seven studies that found balance improvements with durations ranging from two repetitions of 15 seconds to 30 minutes of static stretching. There were three dynamic stretching studies which reported balance improvements (19, 32, 54)

Once again, we have to differentiate between the effects of prolonged static stretching and briefer durations of stretching. Prolonged static stretching has been shown to adversely affect the spinocerebellar pathways (five x 60 seconds of stretching) (116), and ankle proprioception (six x 40 seconds) (130), whereas other studies using three x 30 seconds of static stretching showed no effect (97) or even improvements (58) on knee position sense (proprioception).

In addition to the importance of the vestibular and proprioceptive afferent and efferent

responses for balance (1, 10) the musculotendinous system needs to react and overcome disruptive torques or perturbation in order to restore the centre of gravity to within the base of support (metastability, 89). While a more compliant musculotendinous system may absorb perturbations over a longer duration, allowing greater sensory (afferent) feedback and efferent postural adjustments, a system that is too slack (compliant) necessitates greater shortening of the muscle prolonging the electromechanical delay (time to take up the slack of a compliant tendon in order to exert force on a bone for movement) (18). Hence, the system may react too slowly.

But you also need to differentiate between passive and active stiffness. While prolonged stretching-induced decreases in passive musculotendinous unit are reported, active musculotendinous stiffness may not concomitantly be reduced (18). Thus, while the initial absorption of a perturbation (e.g., a slip or trip) may be more reliant on passive stiffness/compliance, the reaction with an active muscular contraction would be much stiffer accelerating the muscular response. A good example would be the extreme passive compliance of a gymnast or a figure skater to achieve extraordinary feats of flexibility and balance but they can still explode off the floor or ice with jumps, flips, and tumbles due to the active contractions of agonist and antagonist muscles (active stiffness). Furthermore, as stretch training can increase force output at longer muscle lengths, an individual losing their balance may move a leg further to expand the base of support while reacting more forcefully to counterbalance the perturbation (18).

Effects of Flexibility Training

Are the effects of stretching mitigated by population age, sex, trained state or other considerations? Perhaps the reason that so many studies report static stretch-induced deficits is due to the fact that most people do not stretch on a regular basis. If we took a group of untrained people and had them resistance train or go for a long run, they would certainly be sore and weak for days afterwards. Untrained people would experience deficits even with mild to moderate strength or aerobic training. So, would people who train for stretching regularly not experience these stretch-induced deficits? A study from our laboratory reported that 5 weeks of stretch training improved ROM by 12–20 percent but the participants still experienced deficits of 6–8 percent in leg strength and jump height (13). In the same study, we tested a diverse group of people and ranked them by their flexibility. We thought that perhaps the most flexible people might not be as adversely affected by prolonged static stretching. We performed a cross-sectional correlation and did not find any significant relationship between hip and ankle ROM with static stretch-induced deficits in force and jump height. Furthermore, the Behm and Chaouachi review (15) examined 41 stretching studies with trained individuals versus 68 studies that use untrained individuals. They did not find any significant difference overall between the two groups with both experiencing stretch-induced deficits. Prolonged static stretching shows no favourites, trained and untrained, flexible or inflexible people are still susceptible to the adverse effects.

Ageing Effects

There are very few studies examining the effects of stretching on middle aged and older populations. The reports are contradictory with some studies reporting no impairments (71) and others showing impairments (17). Thus, it seems that while the results from individual studies can be contradictory, in general all segments of the population from young to old may experience static-stretch induced impairments from prolonged static stretching that is not accompanied by the dynamic active components of a full warm-up.

Effect of Stretching Intensity

The literature emphasizes the importance of stretch duration for determining whether performance deficits will be incurred (12, 15, 87). These reviews agree that less than 60 seconds of static stretching per muscle group should be implemented if static stretch-induced impairments are to be minimized. What about the intensity of stretching? If some individual stretches to the point of maximal discomfort or pain is it more likely to lead to performance deficits than if they performed lower intensity stretching? Is it "more pain, less gain" for performance after stretching? Once again, the answer is not clear from individual studies. There are some studies showing no impairments when the stretch was held to a point of "mild" discomfort (90, 91, 103, 150), which are contradicted by others that do show performance decrements with stretching to a point of mild discomfort (24, 37, 38, 78, 125). In a study from our laboratory, participants stretched the quadriceps, hamstrings and plantar flexors for 3 repetitions of 30 seconds each either to100 percent, 75 percent or 50 percent of point of maximal discomfort. All stretching intensities adversely affected jump height. To be cautious, individuals should expect performance deficits if they stretch at any intensity for more than 60 seconds per muscle group without the inclusion of any other dynamic activity within the warm-up.

Cost Benefit Analysis

To consider how much flexibility is needed by an individual, a subjective cost-benefit analysis should be conducted. The possible **COST** of prolonged static stretching without a full warm-up is performance impairments of 4–8 percent. The possible **BENEFITS** of static stretching are a greater ROM, more shock absorption, and lower incidence of musculotendinous injuries. A stiff musculotendinous system does not absorb forces well and is more susceptible to strains and sprains. Furthermore, how much ROM is necessary to achieve your goals. For a sprinter, soccer rugby, or ice hockey player, and other sportspeople, a moderate degree of flexibility is necessary. A gymnast, dancer, figure skater, competitive diver, ice hockey goalie, and others need more extreme ROM. To the vast majority of the population who are recreationally active, the cost of possible minor impairments pales in comparison to the cost of injury. A reasonable individual would probably surmise that hitting the golf ball ten yards or metres shorter is a reasonable price to pay to ensure that he/she can continue playing golf multiple times a week throughout the season without injury.

Effects of Dynamic Stretching on Performance

In contrast to static and PNF stretching, dynamic stretching is reported to produce (mean of 48 studies) trivial to small (1.3 percent) performance enhancements (12). The Canadian Society for Exercise Physiology position stand and meta-analysis illustrated that dynamic stretching improved performance in 20 studies (small or greater than small effect sizes), showed trivial effects in 21 studies and impairments in seven studies (12). Measures included countermovement jumps, sprints, agility, isometric and isokinetic force or torque and power, 1RM, balance and other variables. Another review by two

French researchers, Opplert and Babault (113) indicated that there is substantial evidence highlighting the positive effects of dynamic stretching on ROM and subsequent performance (force, power, sprint and jump). They suggested that the greater ROM could be primarily attributed to muscle–tendon unit stiffness reductions, whereas enhanced performance was accredited to elevated temperature and post-activation potentiation enhancement-related mechanisms related to the voluntary contractions of the dynamic stretching.

The average reader should always be cognizant that even when a textbook or a review reports overall significant or meaningful improvements or impairments, a close inspection of the individual studies in the literature almost always shows a spectrum of results that are influenced by the protocol, subject/participant population, the type of measures or other variables. Thus, general summaries or recommendations can be made on the global review of the literature, but there is always the possibility for each individual that their response to a stressor like stretching might not behave similarly to the grand aggregate means reported in these books and reviews.

The review by Behm and Chaouachi (15) recommended that more 90 seconds of dynamic stretching should be performed for each muscle group, but the studies are quite variable in their durations, so a dose response relationship for dynamic stretching is more difficult to establish (12). In these studies, the durations of dynamic stretching can range from one to 20 minutes with either some performance enhancement or no adverse effects. Similarly, there seems to be a tendency for performance improvements with faster dynamic stretches or more intense ballistic stretches, but there is a high degree of variability in the literature, so definitive statements regarding dynamic stretching duration or frequency (speed) also cannot be made at this time. According to one study, the dynamic stretching should be performed before the general warm-up (low-intensity aerobic speeds for an average of five minutes at an individual pace, followed by two x 10 metres sprint) to achieve the best sprint times (70). Another review by our very productive Japanese colleagues (148) recommended that dynamic stretching be conducted "as fast as possible". They also found that explosive performance may be impaired as the dynamic stretching volume increases and thus their regression analysis indicated that the optimal repetitions or distance and sets was "10–15 repetitions" or "10 yards−20 metres" x "1–2 sets", respectively (148). So, if dynamic activities and stretching tend to excite the central nervous system (CNS) whereas static stretching tends to depress the CNS, would a combination of dynamic and static stretching within a warm-up counterbalance each other?

Effects of Combining Static and Dynamic Stretching on Performance

As reported earlier, the studies by Judge and colleagues reported that 39–45 percent of American college coaches for tennis, and track and field use a combination of static and dynamic stretching in their warm-up. Unfortunately, the real-world practices are not paralleled by the volume of research in this area. Thus, even with the scant research combining dynamic and static stretching, can we say whether the neural excitation of the dynamic stretching counteracts the neural inhibition/disfacilitation of the static stretching? Not always! The research results vary from impairments to no change to improvements. In one study, the combination of static and dynamic stretching resulted in 50-metre sprint time impairments compared to dynamic stretching alone (55). Another study showed no difference and no vertical jump impairments when comparing dynamic stretching to a combination of static and dynamic stretching (142). But maybe Gaelic footballers are unique, as in a study using Irish athletes, 20 and 40-metre sprint speed (0.7–1.0 percent) and countermovement jump height (8.7 percent) improved with a combination of static and dynamic stretching (100). A study from our lab, used five conditions 1) general aerobic warm-up with static stretching; 2) general aerobic warmup with dynamic stretching; 3) general and sports specific warm-up with static stretching; and 4) general and sports specific warm-up with dynamic stretching. We found that when a sport specific warm-up was included, there was a 1 percent improvement ($p = 0.001$) in 20-metre sprint time with both the dynamic and static stretch groups. Without a sports specific warm-up, there were no such improvements. In addition, static stretching increased sit and reach (hamstrings and lower back) ROM approximately 3 percent more than the dynamic stretching. In another study, the combination of static and dynamic stretching was not detrimental for the agility of professional soccer players, however dynamic stretching was more effective for preparing for agility (88).

Is performance affected if the individual stretches between sets of a resistance training workout? A study that compared heavy load resistance training (three sets x four repetitions) with heavy load resistance training with stretching (three x 15 seconds) between sets reported no effect of inter-set stretching on jump height (62).

As the research recommends less than 60 seconds of static stretching per muscle group so as not to induce impairments and that many of the ROM studies favour static stretching over dynamic stretching for the greatest ROM increases, it would seem reasonable to recommend a combination of brief (≤60 seconds) static stretching with more than 90 seconds per muscle group of dynamic stretching in addition to a prior ≥five minutes of aerobic activity and a subsequent 5–15 minutes of activity specific dynamic activity to ensure an optimal ROM with either no deficits or perhaps even enhancement of performance. This type of warm-up has been reported in the literature to increase ROM without subsequent impairments or in some cases actually improve some aspects of performance (12).

Effects of Ballistic Stretching on Performance

Ballistic and dynamic stretching are often confused as being the same action. As mentioned before, dynamic stretching involves moving a joint actively through a full ROM under controlled conditions (not maximum speed), whereas ballistic stretching is usually performed at higher speeds often with bouncing movements at the end of the ROM. As usual, the literature is not unanimous. Some articles report that ballistic stretching has no significant effect on subsequent strength (5, 7), vertical jump height (24, 82) or plantar flexors passive resistive torque (102), while others report decreases in knee flexion and extension strength (98, 109), fatigue endurance (7, 98), passive resistive torque and muscle stiffness (95) and Achilles tendon stiffness (102). Fletcher used the term fast dynamic stretching for movements at 100 beats per minute, which might be considered ballistic. With these fast dynamic (maybe ballistic) stretches, they found an increase in vertical jump height and EMG compared to slow dynamic stretches or no stretching at all. Konrad et al.'s (95) multivariate analysis indicated no clinically relevant difference between acute bouts of static, ballistic and PNF stretching. In all these studies, a full warm-up (initial aerobic component, static stretching or slow dynamic stretching followed by sport specific dynamic activity) was not incorporated. Hence, these cited studies should warn us about performing ballistic stretches as the only activity in a warm-up as there is some evidence for performance decrements. The following section describes the problems of interpreting such delimited studies when applying their results to the real world.

Full Warm-ups

The emphasis in the previous section was to warn about the negative effects of "PROLONGED" static and PNF stretching. The typical stretch durations for professional basketball, American football, ice hockey, and baseball players ranges from 12–17 seconds (47, 48, 49, 129). Powerlifters surveyed from 51 countries did static stretches before training for an average of 6.72± 10.31 stretch repetitions for an average duration of 30.8±31.4 seconds. Dynamic stretching was performed by 90.4 percent of the powerlifters (131). However, in the research you see stretch durations of 9–13 repetitions of 135 seconds each (20–30 minutes for the plantar flexors) (36, 56). One of our early studies had participants perform 20 minutes of quadriceps stretching (14). Scientists would say that these studies lack "ecological validity". That is, they are not representative of what actually happens in the real world. Almost nobody would stretch one muscle group for 20–30 minutes especially without all the components of a full warm-up before a training session or competition. Many of the static stretching studies in the literature lack ecological validity (real-world application). As you would imagine most of these prolonged static stretching studies had substantial stretch-induced deficits.

Another major problem is that most studies intervened with a stretching only protocol with no or little other dynamic activity (12). The traditional or conventional pre-activity warm-up protocol consists of an initial submaximal or low intensity exercise component (e.g., running, cycling) that increases the heart rate and muscle temperature (remember the effects of heat on ROM) as well as exciting/activating the neuromuscular system. This aerobic segment is followed by static stretching, which according to our recommendations should not exceed more than 60 seconds per muscle group (12, 15, 87). Dynamic stretching should also be involved to provide more task and velocity specific ROM augmenting activity. The final aspect involves activity or sport specific dynamic activity that further activates the cardiovascular, neuromuscular, endocrine and other physiological systems in preparation for the activity (12, 149). The few studies that do include a full warm-up typically do not illustrate significant stretch-inducing deficits or may provide some performance improvements (99, 108, 124, 134, 143). One of our collaborative studies demonstrated that the combination of prior running, followed by static stretching finishing with countermovement jumps resulted in 7–17 percent better jump performances versus static stretching alone (152). In another collaborative study with our colleagues in Tunisia, the combination of static and dynamic stretching at different static stretching intensities (point of discomfort and less than point of discomfort) resulted in no significant impairments. Thus, the inhibition or depression of the neuromuscular system with prolonged static stretching seems to be balanced by the excitation of the system with dynamic warm-up activities.

A fairly definitive study by our lab (119) demonstrated relatively clearly the interaction of static stretch duration and the dynamic components of the warm-up. In this study, participants went through a full warm-up protocol including a 5-minute aerobic warm-up, followed by a static stretching component, dynamic stretching component, and finally a sport specific dynamic activity component. They either did static stretches of 30, 60, or 120 seconds or no static stretching within this warm-up protocol. Participants were tested after each component as well as ten minutes after the warm-up. All conditions improved hamstrings and quadriceps ROM. There were no impairments in MVC forces, jump height with the control and 30 seconds of static stretching at any time. Impairments were evident with 120 seconds of stretching for quadriceps force (post-static stretching and at ten minutes' post warm-up), force produced in the first 100 milliseconds (post-static stretching), and muscle activation (ten minutes post-warm-up). Vertical jump improved with the warm-up activities except for 60 and 120 seconds of static stretching when tested immediately after the static stretching component. Thus, as the Behm et al. review (12) proposed, static stretching of less than 60 seconds per muscle group within a full warm-up should not lead to performance impairments. The lack of static stretch-induced impairments when incorporated within a full warm-up could be due to the post-activation potentiation effect of the dynamic activities. Kummel et al. (96) reported the typical static stretch deficits but when they added hops, there was no difference from the control condition but the jump heights were significantly higher than the stretch only condition. Similarly, ten minutes of dynamic stretching were sufficient to potentiate vertical jump characteristics (141). Even adding a deadlift did not augment the potentiating effect on vertical jump. Thus, dynamic stretching alone can potentiate performance and offset possible impairments from prolonged static stretching.

These findings were also substantiated by Blazevich and colleagues (21) who incorporated 30 seconds of static stretching into a full warm-up and reported increased ROM but no impairments in jumping, sprinting or agility tests. A very interesting aspect of their study was that they asked the subjects about their performance expectations when stretching was included. Before the study, 18 out of 20 participants nominated dynamic stretching as the most likely to improve performance and 15 out of 20 participants nominated no stretching to be least likely. Nonetheless, these ratings were not related to test performances. The authors suggested that including static or dynamic stretching into a warm-up routine instilled more confidence in the participants regarding their performance in the ensuing sports-related tests. As psychological effects are so potent for optimal performance, and appropriate durations of static and dynamic stretching generally do not seem to degrade performance when incorporated into a full warm-up, it might be a good idea to include short to moderate durations of static stretching (<60 seconds) into a complete warm-up.

Limitations of Acute Stretching Studies

So, athletes, coaches, fitness instructors and enthusiasts have been cautioned by a number of reviews since 2011 about the negative effects of prolonged static stretching especially without comprehensive dynamic warm-up components (12, 15, 30, 87). We published an updated review in 2021 and found 27 acute studies published since the 2016 review, that reported on average a mean moderate magnitude (8.04 percent; d = 0.55) ROM increase and a small magnitude (−1.5 percent; d = 0.36) performance impairment after static stretching (16). Unfortunately, there were still ecological validity problems with only 18 out of 27 (66.6 percent) studies including a full warm-up, 18 out

Table 9.1 Full warm-up components

Submaximal Aerobic Activity	*Static Stretching*	*Dynamic Stretching*	*Task-Specific Activities*
5–15 minutes of running, cycling or other activity to increase muscle temperatures, decrease tissue visco-elasticity, increase heart rate, enzymatic cycling and other factors	<60 seconds per muscle group No need to go to the point of discomfort or pain Stretch major muscle groups and specific muscle groups to the activity	>90 seconds per muscle group Use full range of motion with a controlled movement at moderate speeds	5–15 minutes Practice movements that are associated with the sport or task at velocities close to the actual movement

of 27 (66.6 percent) also implementing more than 60 seconds of static stretching, four out of 25 (14.8 percent) added dynamic stretches or activities after static stretching and only ten out of 27 (37.1 percent) performed testing at more than five minutes after stretching.

After the warm-up, do most athletes or teams immediately begin their competition? Many athletes or teams go back to their dressing room or bench to finalize tactics, strategies, make equipment adjustments, and other necessities (i.e., pre-game renal voiding). For some competitions, there is a national anthem to start the game. But the average time to testing after a warm-up was 3.2, 4.9, and 4.1 minutes for static, dynamic and PNF stretching studies (12). Once again, the ecological validity can be questioned, as many athletes would not start their competition less than 5 minutes after their warm-up.

Furthermore, who are the subjects in these studies? As the majority of the studies are conducted at the Kinesiology, Physical Education, Sport or Exercise Science departments of universities, the students are likely to be familiar with these studies that report stretch-induced impairments. A study from our laboratory tried to deceive the participants by telling non-Kinesiology students that stretching would actually improve their performance and compared their responses to Kinesiology students (knowledgeable about the stretching literature) after 3 static hamstrings stretches of 30 seconds each. Both groups still experienced hamstring force and activation deficits due to stretching but the deceived group experienced improvements in the performance of the antagonist quadriceps muscle. Therefore, the deception could not overcome the stretching impairments of the hamstrings (90 seconds was 50 percent more than the recommended ≤60 seconds) but the quadriceps, which were not stretched, were susceptible to the deception that stretching would improve performance. Hence bias and deception could play a small role in these types of studies. Furthermore, there is bias in the journal publications for studies that provide significant findings (46). For example, it is more difficult to publish a study that reports no changes with stretching than a study that would report significant decrements. Our review found that reporting of data/statistics within published papers was less common when non-significant results were obtained. In most studies the details of non-significant changes were commonly not reported, which could result in an overestimation of the stretching deficit effects (12).

Improving Performance by Limiting Muscle Soreness

Excessive unaccustomed activity especially those activities that emphasize eccentric contractions can lead to exercise-induced muscle damage (EIMD) leading to delayed onset muscle soreness (DOMS) that can persist for 5–7 days (11, 68). EIMD and DOMS can decrease force, power, running speed, neuromuscular efficiency (extent of muscle activation needed to accomplish a task), ROM and many other performance factors (26, 34, 35). It has been a common belief that stretching can either prevent or reduce EIMD and

DOMS. Stretching either before, during or after the EIMD-induced exercise did not reduce subsequent DOMS (25, 68, 75, 76). A review by Howatson and van Someren (79) also concluded that stretching to prevent or reduce EIMD provides only minimal effects in reducing muscle soreness with no performance effects. Some early studies reported that during DOMS, static stretch-induced reductions in muscle activity (measured with EMG or measurement of the H-reflex to muscle action potential ratio) (41, 42, 43, 135) contributed to reduced soreness. For a similar reason (decrease muscle activity) (43, 74), static stretching tends to relieve muscle cramps (20, 40, 144). Other possible suggested mechanisms for DOMS relief would be that stretching helps to remove some of the inflammation or swelling associated with DOMS (22). Furthermore, stretching during DOMS could initiate the Gate Control Theory of Pain (106) suppression whereby the activation of higher velocity afferents (i.e. type Ia, Ib,) from mechanical stimulation of stretch, cutaneous and other receptors can inhibit the slower velocity pain afferents such as type IV afferents (2) (see Figure 9.7). A common example of the Gate Control Theory is the typical response to an injured finger or limb or other body part. The immediate response is to grab that injured body part and start rubbing it. Rubbing the area activates the high velocity cutaneous, tactile or pressure receptors that help to peripherally and centrally block the slower afferents from the nociceptors (pain receptors). Typically, there is never just one mechanism and thus DOMS reduction could be a combination of a number of mechanisms.

Summary

Throughout the second half of the 20th century, it was commonly believed that static stretching-induced increases in muscle compliance (reduced stiffness) would enhance subsequent performance by allowing a greater ROM with less resistance to movement. However, static and PNF stretching studies from the late 1990s into the early 2000s predominantly reported subsequent performance impairments of 3–7.5 percent. These findings changed the traditional warm-up paradigm shifting the emphasis from static to dynamic stretching within a warm-up. Specifically, it was found that prolonged static stretching (>60 seconds per muscle group) without all components of a full warm-up increased the chances for performance deficits. The decision to include static stretching within a warm-up may depend upon the type or demands of the activity (i.e., jogging with minimal ROM needs versus hurdling with increased flexibility needs), the level of the athlete (elite vs. recreational) and the priority of reducing the incidence of musculotendinous injuries compared to possible 3–4 percent performance decrements. Prolonged static stretching without a full warm-up can impair performance in both trained and untrained individuals, young and old, whether the stretch is performed at a moderate or high intensity.

Dynamic stretching, for its part, may have either no significant impairments or actually provide small enhancements of subsequent performance.

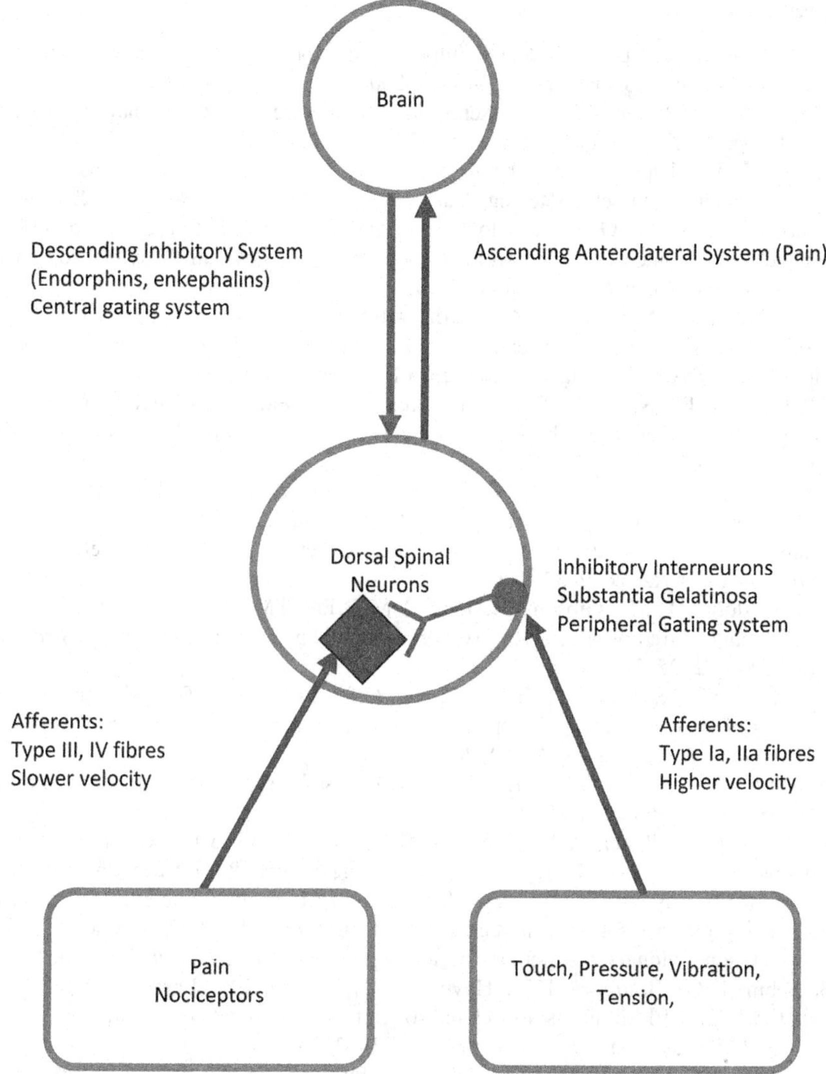

Figure 9.7 Gate Control Theory of Pain (Melzack and Wall)

Adding static to dynamic stretching tends to counterbalance possible impairments with prolonged static stretching. Individuals should incorporate at least 90 seconds of dynamic stretching per muscle group within a full warm-up. The full warm-up should include a 5–15-minute submaximal aerobic component, less than 60 seconds of static stretching per muscle group, more than 90 seconds of dynamic stretching per muscle group and finish with dynamic activity that should be activity or sport specific.

References

1. Anderson, K. and Behm, D.G.The impact of instability resistance training on balance and stability. *Sports Med* 35: 43–53, 2005.
2. Armstrong, R.B.Mechanisms of exercise-induced delayed onset muscular soreness: a brief review. *Med Sci Sports Exerc* 16: 529–538, 1984.
3. Avela, J., Kyröläinen, H., and Komi, P.V.Altered reflex sensitivity after repeated and prolonged passive muscle stretching. *Journal of Applied Physiology* 86: 1283–1291, 1999.
4. Babault, N., Rodot, G., Champelovier, M., and Cometti, C.A Survey on Stretching Practices in Women and Men from Various Sports or Physical Activity Programs. *Int J Environ Res Public Health* 18, 2021.
5. Bacurau, R.F., Monteiro, G.A., Ugrinowitsch, C., Tricoli, V., Cabral, L.F., and Aoki, M.S.Acute effect of a ballistic and a static stretching exercise bout on flexibility and maximal strength. *J Strength Cond Res* 23: 304–308, 2009.
6. Baptista, R.R., Scheeren, E.M., Macintosh, B.R., and Vaz, M.A.Low-frequency fatigue at maximal and submaximal muscle contractions. *Braz J Med Biol Res* 42: 380–385, 2009.
7. Barroso, R., Tricoli, V., Santos Gil, S.D., Ugrinowitsch, C., and Roschel, H.Maximal strength, number of repetitions, and total volume are differently affected by static-, ballistic-, and proprioceptive neuromuscular facilitation stretching. *J Strength Cond Res* 26: 2432–2437, 2012.
8. Bazett-Jones, D.M., Gibson, M.H., and McBride, J.M.Sprint and vertical jump performances are not affected by six weeks of static hamstring stretching. *J Strength Cond Res* 22: 25–31, 2008.
9. Beckett, J.R., Schneiker, K.T., Wallman, K.E., Dawson, B.T., and Guelfi, K.J. Effects of static stretching on repeated sprint and change of direction performance. *MedSciSports Exerc* 41: 444–450, 2009.
10. Behm, D.G. and Anderson, K.G.The role of instability with resistance training. *J Strength Cond Res* 20: 716–722, 2006.
11. Behm, D.G., Baker, K.M., Kelland, R., and Lomond, J.The effect of muscle damage on strength and fatigue deficits. *J Strength Cond Res* 15: 255–263, 2001.
12. Behm, D.G., Blazevich, A.J., Kay, A.D., and McHugh, M.Acute effects of muscle stretching on physical performance, range of motion, and injury incidence in healthy active individuals: a systematic review. *Appl Physiol Nutr Metab* 41: 1–11, 2016.
13. Behm, D.G., Bradbury, E.E., Haynes, A.T., Hodder, J.N., Leonard, A.M., and Paddock, N.R.Flexibility is not related to stretch-induced deficits in force or power. *Journal of Sports Science and Medicine* 5: 33–42, 2006.
14. Behm, D.G., Button, D.C., and Butt, J.C.Factors affecting force loss with prolonged stretching. *Can J Appl Physiol* 26: 261–272, 2001.
15. Behm, D.G. and Chaouachi, A.A review of the acute effects of static and dynamic stretching on performance. *Eur J Appl Physiol* 111: 2633–2651, 2011.
16. Behm, D.G., Kay, A.D., Trajano, G.S., and Blazevich, A.J.Mechanisms underlying performance impairments following prolonged static stretching without a comprehensive warm-up. *Eur J Appl Physiol* 121: 67–94, 2021.
17. Behm, D.G., Plewe, S., Grage, P., Rabbani, A., Beigi, H.T., Byrne, J.M., and Button, D.C.Relative static stretch-induced impairments and dynamic stretch-induced enhancements are similar in young and middle-aged men. *Appl Physiol Nutr Metab* 36: 790–797, 2011.

18. Behm, D.G., Kay, A.D., Trajano, G.S., and Alizadeh, S., and Blazevich, A.J. Effects of stretching on injury risk reduction and balance. *Journal of Clinical Exercise Physiology* 10: 106–116, 2021.
19. Belkhiria-Turki, L., Chaouachi, A., Turki, O., Hammami, R., Chtara, M., Amri, M., Drinkwater, E.J., and Behm, D.G.Greater volumes of static and dynamic stretching within a warm-up do not impair star excursion balance performance. *J Sports Med Phys Fitness* 54: 279–288, 2014.
20. Bertolasi, L., De Grandis, D., Bongiovanni, L.G., Zanette, G.P., and Gasperini, M. The influence of muscular lengthening on cramps. *Ann Neurol* 33: 176–180, 1993.
21. Blazevich, A.J., Gill, N.D., Kvorning, T., Kay, A.D., Goh, A., Hilton, B., Drinkwater, E.J., and Behm, D.G.No Effect of Muscle Stretching within a Full, Dynamic Warm-up on Athletic Performance. *Med Sci Sports Exerc*, 2018.
22. Bobbert, M.F., Hollander, A.P., and Huijing, P.A.Factors in delayed onset muscular soreness of man. *Med Sci Sports Exerc* 18: 75–81, 1986.
23. Borges Bastos, C.L., Miranda, H., Vale, R.G., Portal Mde, N., Gomes, M.T., Novaes Jda, S., and Winchester, J.B.Chronic effect of static stretching on strength performance and basal serum IGF-1 levels. *J Strength Cond Res* 27: 2465–2472, 2013.
24. Bradley, P.S., Olsen, P.D., and Portas, M.D.The effect of static, ballistic, and proprioceptive neuromuscular facilitation stretching on vertical jump performance. *J Strength Cond Res* 21: 223–226, 2007.
25. Buroker, K.C. and Schwane, J.A.Does Postexercise Static Stretching Alleviate Delayed Muscle Soreness? *Phys Sportsmed* 17: 65–83, 1989.
26. Byrnes, W.C. and Clarkson, P.M.Delayed onset muscle soreness and training. *Clinics in Sports Medicine* 5: 605–615, 1986.
27. Cappa, D.F. and Behm, D.G.Training Specificity of Hurdle vs. Countermovement Jump Training. *Journal of Strength and Conditioning Research/National Strength & Conditioning Association*, 2011.
28. Cavagna, G.A., Dusman, B., and Margaria, R.Positive work done by a previously stretched muscle. *J Appl Physiol* 24: 21–32, 1968.
29. Ce, E., Paracchino, E., and Esposito, F.Electrical and mechanical response of skeletal muscle to electrical stimulation after acute passive stretching in humans: a combined electromyographic and mechanomyographic approach. *J Sports Sci* 26: 1567–1577, 2008.
30. Chaabene, H., Behm, D.G., Negra, Y., and Granacher, U.Acute Effects of Static Stretching on Muscle Strength and Power: An Attempt to Clarify Previous Caveats. *Front Physiol* 10: 1468, 2019.
31. Chaouachi, A., Chamari, K., Wong, P., Castagna, C., Chaouachi, M., Moussa-Chamari, I., and Behm, D.G.Stretch and sprint training reduces stretch-induced sprint performance deficits in 13- to 15-year-old youth. *Eur J Appl Physiol* 104: 515–522, 2008.
32. Chatzopoulos, D., Galazoulas, C., Patikas, D., and Kotzamanidis, C.Acute Effects of Static and Dynamic Stretching on Balance, Agility, Reaction Time and Movement Time. *Journal of Sports Science and Medicine* 13: 403–409, 2014.
33. Church, J.B., Wiggins, M.S., Moode, E.M., and Crist, R.Effect of warm-up and flexibility treatments on vertical jump performance. *Journal of Strength and Conditioning Research* 15: 332–336, 2001.
34. Cleak, M.J. and Eston, R.G.Delayed onset muscle soreness: Mechanisms and management. *Journal of Sports Sciences* 10: 325–341, 1992.

35. Connolly, D.A., Sayers, S.P., and McHugh, M.P.Treatment and prevention of delayed onset muscle soreness. *J Strength Cond Res* 17: 197–208, 2003.
36. Costa, P.B., Ryan, E.D., Herda, T.J., Walter, A.A., Hoge, K.M., and Cramer, J.T. Acute effects of passive stretching on the electromechanical delay and evoked twitch properties. *European Journal of Applied Physiology and Occupational Physiology* 108: 301–310, 2010.
37. Cramer, J.T., Housh, T.J., Johnson, G.O., Miller, J.M., Coburn, J.W., and Beck, T. W.Acute effects of static stretching on peak torque in women. *Journal of Strength and Conditioning Research* 18: 236–241, 2004.
38. Cramer, J.T., Housh, T.J., Weir, J.P., Johnson, G.O., Coburn, J.W., and Beck, T.W. The acute effects of static stretching on peak torque, mean power output, electromyography, and mechanomyography. *European Journal of Applied Physiology* 93: 530–539, 2005.
39. Cronin, J., McNair, P.J., and Marshall, R.N.Power absorption and production during slow, large-amplitude stretch-shorten cycle motions. *Eur J Appl Physiol* 87: 59–65, 2002.
40. Davison, S.Standing: a good remedy. *JAMA* 252: 3367, 1984.
41. de Vries, H.A.Electromyographic observation of the effect of static stretching upon muscular distress. *Research Quarterly* 4: 468–479, 1961.
42. de Vries, H.A.Prevention of muscular distress after exercise. *Research Quarterly* 2: 177–185, 1961b.
43. de Vries, H.A.Quantitative electromyographic investigation of the spasm theory of muscle pain. *Am J Phys Med* 45: 119–134, 1966.
44. Dietz, V., Schmidtblecher, D., and Noth, J.Neuronal Mechanisms of Human Locomotion. *Journal Of Neurophysiology* 42: 1212–1222, 1979.
45. Dintiman, G.B.Effects of Various Training Programs on Running Speed. *Res Q* 35: 456–463, 1964.
46. Easterbrook, P.J., Berlin, J.A., Gopalan, R., and Matthews, D.R.Publication bias in clinical research. *Lancet* 337: 867–872, 1991.
47. Ebben, W.P. and Blackard, D.O.Strength and conditioning practices of National Football League strength and conditioning coaches. *J Strength Cond Res* 15: 48–58, 2001.
48. Ebben, W.P., Carroll, R.M., and Simenz, C.J.Strength and conditioning practices of National Hockey League strength and conditioning coaches. *J Strength Cond Res* 18: 889–897, 2004.
49. Ebben, W.P., Hintz, M.J., and Simenz, C.J.Strength and conditioning practices of Major League Baseball strength and conditioning coaches. *J Strength Cond Res* 19: 538–546, 2005.
50. ElBasiouny, S.M., Schuster, J.E., and Heckman, C.J.Persistent inward currents in spinal motoneurons: important for normal function but potentially harmful after spinal cord injury and in amyotrophic lateral sclerosis. *Clin Neurophysiol* 121: 1669–1679, 2010.
51. Esposito, F., Ce, E., Rampichini, S., and Veicsteinas, A.Acute passive stretching in a previously fatigued muscle: Electrical and mechanical response during tetanic stimulation. *J Sports Sci* 27: 1347–1357, 2009.
52. Esposito, F., Limonta, E., and Ce, E.Passive stretching effects on electromechanical delay and time course of recovery in human skeletal muscle: new insights from an electromyographic and mechanomyographic combined approach. *European Journal of Applied Physiology* 111: 485–495, 2011.

53. Etnyre, B.R. and Abraham, L.D.H-reflex changes during static stretching and two variations of proprioceptive neuromuscular facilitation techniques. *Electro-encephalography Clinical Neurophysiology* 63: 174–179, 2005.
54. Fletcher, I.M.The effect of different dynamic stretch velocities on jump performance. *Eur J Appl Physiol* 109: 491–498, 2010.
55. Fletcher, I.M. and Anness, R.The acute effects of combined static and dynamic stretch protocols on fifty-meter sprint performance in track-and-field athletes. *J Strength Cond Res* 21: 784–787, 2007.
56. Fowles, J.R., Sale, D.G., and MacDougall, J.D.Reduced strength after passive stretch of the human plantar flexors. *Journal of applied physiology* 89: 1179–1188, 2000.
57. Gandevia, S.C.Neural control in human muscle fatigue: changes in muscle afferents, moto neurones and moto cortical drive. *Acta Physiologica Scandinavica* 162: 275–283, 1998.
58. Ghaffarinejad, F., Taghizadeh, S., and Mohammadi, F.Effect of static stretching of muscles surrounding the knee on knee joint position sense. *Br J Sports Med* 41: 684–687, 2007.
59. Gleim, G.W., Stachenfeld, N.S., and Nicholas, J.A.The influence of flexibility on the economy of walking and jogging. *J Orthop Res* 8: 814–823, 1990.
60. Godges, J.J., MacRae, H., Longdon, C., and Tinberg, C.The effects of two stretching procedures on the economy of walking and jogging. *Journal of Orthopaedics and Sport Physical Therapy* 7: 350–357, 1989.
61. Gollhofer, A., Strojnik, V., Rapp, W., and Schweizer, L.Behaviour of triceps surae muscle-tendon complex in different jump conditions. *Eur J Appl Physiol Occup Physiol* 64: 283–291, 1992.
62. Gonzalez-Rave, J.M., Machado, L., Navarro-Valdivielso, F., and Vilas-Boas, J.P. Acute effects of heavy-load exercises, stretching exercises, and heavy-load plus stretching exercises on squat jump and countermovement jump performance. *J Strength Cond Res* 23: 472–479, 2009.
63. Grosset, J.F., Piscione, J., Lambertz, D., and Perot, C.Paired changes in electromechanical delay and musculo-tendinous stiffness after endurance or plyometric training. *European Journal of Applied Physiology and Occupational Physiology* 105: 131–139, 2009.
64. Guissard, N. and Duchateau, J.Neural aspects of muscle stretching. *ExercSport SciRev* 34: 154–158, 2006.
65. Guissard, N., Duchateau, J., and Hainaut, K.Muscle stretching and motoneuron excitability. *European Journal of Applied Physiology* 58: 47–52, 1988.
66. Guissard, N., Duchateau, J., and Hainaut, K.Mechanisms of decreased motoneuron excitation during passive muscle stretching. *Experimental Brain Research* 137: 163–169, 2001.
67. Gulati, A., Jain, R., Lehri, A., and Kumar, R.Effect of high and low flexibility on agility, acceleration speed and vertical jump performance of volleyball players. *European Journal of Physical Education and Sport Science* 6: 120–130, 2021.
68. Gulick, D.T., Kimura, I.F., Sitler, M., Paolone, A., and Kelly, J.D.Various treatment techniques on signs and symptoms of delayed onset muscle soreness. *JAthl-Train* 31: 145–152, 1996.
69. Gultekin, Z., Kin-Isler, A., and Surenkok, O.Hemodynamic and lactic acid responses to proprioceptive neuromuscular facilitation exercise. *Journal of Sports Science and Medicine* 5: 375–380, 2006.

70. Gunay, E., Azcelik, R., Manci, E., Cetinkaya, C., and Bediz, C.S.For a Higher Sprint Running Performance, in Which Part of the Warm-Up Protocol Should the Dynamic Stretching Phase be Applied? *Turkish Journal of Sport and Exercise* 25: 27–34, 2023.
71. Handrakis, J.P., Southard, V.N., Abreu, J.M., Aloisa, M., Doyen, M.R., Echevarria, L.M., Hwang, H., Samuels, C., Venegas, S.A., and Douris, P.C.Static Stretching Does Not Impair Performance in Active Middle-Aged Adults. *J Strength Cond Res* 24: 825–830, 2010.
72. Hayes, P.R. and Walker, A.Pre-exercise stretching does not impact upon running economy. *J Strength Cond Res* 21: 1227–1232, 2007.
73. Hayes, W.C., Myers, E.R., Morris, J.N., Gerhart, T.N., Yett, H.S., and Lipsitz, L. A.Impact near the hip dominates fracture risk in elderly nursing home residents who fall. *Calcif Tissue Int* 52: 192–198, 1993.
74. Helin, P.Physiotherapy and electromyography in muscle cramp. *Br J Sports Med* 19: 230–231, 1985.
75. Herbert, R.D. and Gabriel, M.Effects of stretching before and after exercising on muscle soreness and risk of injury: systematic review. *BMJ* 325: 468, 2002.
76. High, D.M., Howley, E.T., and Franks, B.D.The effects of static stretching and warm-up on prevention of delayed-onset muscle soreness. *Res Q Exerc Sport* 60: 357–361, 1989.
77. Hoch, M.C., Staton, G.S., and McKeon, P.O.Dorsiflexion range of motion significantly influences dynamic balance. *J Sci Med Sport* 14: 90–92, 2011.
78. Hough, P.A., Ross, E.Z., and Howatson, G.Effects of dynamic and static stretching on vertical jump performance and electromyographic activity. *J Strength Cond Res* 23: 507–512, 2009.
79. Howatson, G. and van Someren, K.A.The prevention and treatment of exercise-induced muscle damage. *Sports Med* 38: 483–503, 2008.
80. Ishikawa, M., Komi, P.V., Finni, T., and Kuitunen, S.Contribution of the tendinous tissue to force enhancement during stretch-shortening cycle exercise depends on the prestretch and concentric phase intensities. *J Electromyogr Kinesiol* 16: 423–431, 2006.
81. Jager, T.E., Weiss, H.B., Coben, J.H., and Pepe, P.E.Traumatic brain injuries evaluated in U.S. emergency departments, 1992–1994. *Acad Emerg Med* 7: 134–140, 2000.
82. Jaggers, J.R., Swank, A.M., Frost, K.L., and Lee, C.D.The acute effects of dynamic and ballistic stretching on vertical jump height, force, and power. *J Strength Cond Res* 22: 1844–1849, 2008.
83. Jones, A.M.Running economy is negatively related to sit-and-reach test performance in international-standard distance runners. *International Journal of Sports Medicine* 23: 40–43, 2002.
84. Judge, L.W., Bellar, D., Craig, B., Petersen, J., Camerota, J., Wanless, E., and Bodey, K.An examination of preactivity and postactivity flexibility practices of National Collegiate Athletic Association Division I tennis coaches. *J Strength Cond Res* 26: 184–191, 2012.
85. Judge, L.W., Bellar, D.M., Gilreath, E.L., Petersen, J.C., Craig, B.W., Popp, J.K., Hindawi, O.S., and Simon, L.S.An examination of preactivity and postactivity stretching practices of NCAA division I, NCAA division II, and NCAA division III track and field throws programs. *J Strength Cond Res* 27: 2691–2699, 2013.
86. Judge, L.W., Petersen, J.C., Bellar, D.M., Craig, B.W., Wanless, E.A., Benner, M., and Simon, L.S.An examination of preactivity and postactivity stretching practices of cross-country and track and field distance coaches. *J Strength Cond Res* 27: 2456–2464, 2013.

87. Kay, A.D. and Blazevich, A.J.Effect of acute static stretch on maximal muscle performance: a systematic review. *Med Sci Sports Exerc* 44: 154–164, 2012.
88. Khorasani, M.A., Sahebozomai, M., Tabrizi, K.G., and Yusof, A.B.Acute effect of different stretching methods on Illinois agility test in soccer players. *Journal of Strength and Conditioning Research*: 1–7, 2010.
89. Kibele, A., Granacher, U., Muehlbauer, T., and Behm, D.G.Stable, Unstable and Metastable States of Equilibrium: Definitions and Applications to Human Movement. *J Sports Sci Med* 14: 885–887, 2015.
90. Knudson, D., Bennett, K., Corn, R., Leick, D., and Smith, C.Acute effects of stretching are not evident in the kinematics of the vertical jump. *Journal of Strength and Conditioning Research* 15: 98–101, 2001.
91. Knudson, D.V., Noffal, G.J., Bahamonde, R.E., Bauer, J.A., and Blackwell, J. R.Stretching has no effect on tennis serve performance. *J Strength Cond Res* 18: 654–656, 2004.
92. Kokkonen, J., Nelson, A.G., and Cornwell, A.Acute muscle stretching inhibits maximal strength performance. *Research Quarterly for Exercise and Sport* 69: 411–415, 1998.
93. Komi, P.V. *The Stretch-shortening Cycle and Human Power Output*. Champaign, IL: Human Kinetics, 1986.
94. Komi, P.V. and Bosco, C.Utilization of stored elastic energy in leg extensor muscles by men and women. *Medicine and Science in Sports* 10: 261–265, 1978.
95. Konrad, A., Stafilidis, S., and Tilp, M.Effects of acute static, ballistic, and PNF stretching exercise on the muscle and tendon tissue properties. *Scand J Med Sci Sports* 27: 1070–1080, 2017.
96. Kummel, J., Kramer, A., Cronin, N.J., and Gruber, M.Postactivation potentiation can counteract declines in force and power that occur after stretching. *Scand J Med Sci Sports*, 2016.
97. Larsen, R., Lund, H., Christensen, R., Rogind, H., Danneskiold-Samsoe, B., and Bliddal, H.Effect of static stretching of quadriceps and hamstring muscles on knee joint position sense. *Br J Sports Med* 39: 43–46, 2005.
98. Lima, C.D., Brown, L.E., Wong, M.A., Leyva, W.D., Pinto, R.S., Cadore, E.L., and Ruas, C.V.Acute Effects of Static vs. Ballistic Stretching on Strength and Muscular Fatigue Between Ballet Dancers and Resistance-Trained Women. *J Strength Cond Res* 30: 3220–3227, 2016.
99. Little, T. and Williams, A.G.Effects of Differential Stretching Protocols During Warm-Ups on High-Speed Motor Capacities In Professional Soccer Players. *Journal of Strength and Conditioning Research* 20: 203–207, 2006.
100. Loughran, M., Glasgow, P., Bleakley, C., and McVeigh, J.The effects of a combined static-dynamic stretching protocol on athletic performance in elite Gaelic footballers: A randomised controlled crossover trial. *Phys Ther Sport* 25: 47–54, 2017.
101. Maddigan, M.E., Peach, A.A., and Behm, D.G.A comparison of assisted and unassisted proprioceptive neuromuscular facilitation techniques and static stretching. *Journal of Strength and Conditioning Research/National Strength & Conditioning Association* 26: 1238–1244, 2012.
102. Mahieu, N.N., McNair, P., De Muynck, M., Stevens, V., Blanckaert, I., Smits, N., and Witvrouw, E.Effect of static and ballistic stretching on the muscle-tendon tissue properties. *Med Sci Sports Exerc* 39: 494–501, 2007.

103. Manoel, M.E., Harris-Love, M.O., Danoff, J.V., and Miller, T.A.Acute effects of static, dynamic, and proprioceptive neuromuscular facilitation stretching on muscle power in women. *J Strength Cond Res* 22: 1528–1534, 2008.
104. Marek, S.M., Cramer, J.T., Fincher, A.L., Massey, L.L., Dangelmaier, S.M., Purkayastha, S., Fitz, K.A., and Culbertson, J.Y.Acute effects of static and proprioceptive neuromuscular facilitation stretching on muscle strength and power output. *Journal of Athletic Training* 40: 94–103, 2005.
105. McArdle, W.D., Katch, F.I., and Katch, V.L. *Exercise Physiology: Energy, Nutrition, and Human Performance.* Malvern, PA: Lea and Febiger, 1991.
106. Melzack, R. and Wall, P.D.Pain mechanisms: a new theory. *Science* 150: 971–979, 1965.
107. Mero, A. and Komi, P.V.Reaction time and electromyographic activity during a sprint start. *Eur J Appl Physiol Occup Physiol* 61: 73–80, 1990.
108. Murphy, J.R., DiSanto, M.C., Alkanani, T., and Behm, D.G.Aerobic activity before and following short-duration static stretching improves range of motion and performance vs. a traditional warm-up. *Appl Physiol Nutr Metab* 35: 679–690, 2010.
109. Nelson, A.G. and Kokkonen, J.Acute ballistic muscle stretching inhibits maximal strength performance. *Res Q Exerc Sport* 72: 415–419, 2001.
110. Nelson, A.G., Kokkonen, J., Eldredge, C., Cornwell, A., and Glickman-Weiss, E. Chronic stretching and running economy. *Scand J Med Sci Sports* 11: 260–265, 2001.
111. Nicol, C. and Komi, P.V.Significance of passively induced stretch reflexes on achilles tendon force enhancement. *Muscle Nerve* 21: 1546–1548, 1998.
112. Norman, R.W. and Komi, P.V.Electromechanical delay in skeletal muscle under normal movement conditions. *Acta Physiologica Scandinavica* 106: 241–248, 1979.
113. Opplert, J. and Babault, N.Acute Effects of Dynamic Stretching on Muscle Flexibility and Performance: An Analysis of the Current Literature. *Sports Med* 48: 299–325, 2018.
114. Power, K., Behm, D., Cahill, F., Carroll, M., and Young, W.An acute bout of static stretching: effects on force and jumping performance. *Med Sci Sports Exerc* 36: 1389–1396, 2004.
115. Pulverenti, T.S., Trajano, G.S., Kirk, B.J.C., and Blazevich, A.J.The loss of muscle force production after muscle stretching is not accompanied by altered corticospinal excitability. *Eur J Appl Physiol* 119: 2287–2299, 2019.
116. Pulverenti, T.S., Trajano, G.S., Kirk, B.J.C., Bochkezanian, V., and Blazevich, A. J.Plantar flexor muscle stretching depresses the soleus late response but not tendon tap reflexes. *Eur J Neurosci*, 2021.
117. Pulverenti, T.S., Trajano, G.S., Walsh, A., Kirk, B.J.C., and Blazevich, A.J.Lack of cortical or Ia-afferent spinal pathway involvement in muscle force loss after passive static stretching. *J Neurophysiol* 123: 1896–1906, 2020.
118. Rack, P.M. and Westbury, D.R.The short range stiffness of active mammalian muscle and its effect on mechanical properties. *J Physiol* 240: 331–350, 1974.
119. Reid, J.C., Greene, R., Young, J.D., Hodgson, D.D., Blazevich, A.J., and Behm, D.G.The effects of different durations of static stretching within a comprehensive warm-up on voluntary and evoked contractile properties. *European Journal of Applied Physiology*, 2018.
120. Rey, E., Padron-Cabo, A., Barcala-Furelos, R., and Mecias-Calvo, M.Effect of High and Low Flexibility Levels on Physical Fitness and Neuromuscular Properties in Professional Soccer Players. *Int J Sports Med* 37: 878–883, 2016.

121. Rubini, E.C., Costa, A.L., and Gomes, P.S.The effects of stretching on strength performance. *Sports Med* 37: 213–224, 2007.
122. Ryan, E.D., Beck, T.W., Herda, T.J., Hull, H.R., Hartman, M.J., Stout, J.R., and Cramer, J.T.Do practical durations of stretching alter muscle strength? A dose-response study. *Med Sci Sports Exerc* 40: 1529–1537, 2008.
123. Ryan, E.E., Rossi, M.D., and Lopez, R.The effects of the contract-relax-antagonist-contract form of proprioceptive neuromuscular facilitation stretching on postural stability. *Journal of strength and conditioning research/National Strength & Conditioning Association* 24: 1888–1894, 2010.
124. Samson, M., Button, D.C., Chaouachi, A., and Behm, D.G.Effects of dynamic and static stretching within general and activity specific warm-up protocols. *J Sports Sci Med* 11: 279–285, 2012.
125. Sayers, A.L., Farley, R.S., Fuller, D.K., Jubenville, C.B., and Caputo, J.L.The effect of static stretching on phases of sprint performance in elite soccer players. *J Strength Cond Res* 22: 1416–1421, 2008.
126. Schmidtbleicher, D.Training for power events. In: *Strength and Power in Sport*. P. V. Komi, ed. Oxford: Blackwell Publishers, 1992, pp. 381–395.
127. Shrier, I.Does stretching improve performance?: a systematic and critical review of the literature. *ClinJSport Med* 14: 267–273, 2004.
128. Siatras, T., Papadopoulos, G., Mameletzi, D.N., Vasilios, G., and Kellis, S.Static and dynamic acute stretching effect on gymnasts' speed in vaulting. *Pediatric Exercise Science* 15: 383–391, 2003.
129. Simenz, C.J., Dugan, C.A., and Ebben, W.P.Strength and conditioning practices of National Basketball Association strength and conditioning coaches. *J Strength CondRes* 19: 495–504, 2005.
130. Smajla, D., Garcia-Ramos, A., Tomazin, K., and Strojnik, V.Selective effect of static stretching, concentric contractions, and a one-leg balance task on ankle motion sense in young and older adults. *Gait Posture* 71: 1–6, 2019.
131. Spence, A.J., Helms, E.R., and McGuigan, M.R.Stretching Practices of International Powerlifting Federation Unequipped Powerlifters. *J Strength Cond Res* 36: 3456–3461, 2022.
132. Streepey, J.W., Mock, M.J., Riskowski, J.L., Vanwye, W.R., Vitvitskiy, B.M., and Mikesky, A.E.Effects of quadriceps and hamstrings proprioceptive neuromuscular facilitation stretching on knee movement sensation. *J Strength Cond Res* 24: 1037–1042, 2010.
133. Suzuki, T., Sugawara, K., Iizuka, Y., and Kubo, K.Effects of static stretching on active muscle stiffness with and without the stretch reflex. *Journal of Physical Fitness and Sports Medicine* 9: 37–41, 2020.
134. Taylor, J.M., Weston, M., and Portas, M.D.The effect of a short practical warm-up protocol on repeated sprint performance. *J Strength Cond Res* 27: 2034–2038, 2013.
135. Thigpen, L.K., Moritani, T., Thiebaud, R., and Hargis, J.L.The acute effects of static stretching on alpha motoneuron excitability. In: *Biomechanics IX-A International series on biomechanics*. D.A. Winter*et al.* (Eds). Champaign, IL: Human Kinetics 1985, pp. 352–357.
136. Torres, E.M., Kraemer, W.J., Vingren, J.L., Volek, J.S., Hatfield, D.L., Spiering, B.A., Ho, J.Y., Fragala, M.S., Thomas, G.A., Anderson, J.M., Hakkinen, K., and Maresh, C.M.Effects of stretching on upper body muscular performance. *Journal of Strength and Conditioning Research* 22: 1279–1285, 2008.

137. Toumi, H., Poumarat, G., Best, T.M., Martin, A., Fairclough, J., and Benjamin, M.Fatigue and muscle-tendon stiffness after stretch-shortening cycle and isometric exercise. *Appl Physiol Nutr Metab* 31: 565–572, 2006.
138. Trajano, G.S. and Blazevich, A.J.Static Stretching Reduces Motoneuron Excitability: The Potential Role of Neuromodulation. *Exerc Sport Sci Rev* 49: 126–132, 2021.
139. Trajano, G.S., Nosaka, K., and Blazevich, A.J.Neurophysiological Mechanisms Underpinning Stretch-Induced Force Loss. *Sports Med* 47: 1531–1541, 2017.
140. Trehearn, T.L. and Buresh, R.J.Sit-and-reach flexibility and running economy of men and women collegiate distance runners. *J Strength Cond Res* 23: 158–162, 2009.
141. Turki, O., Chaouachi, A., Drinkwater, E.J., Chtara, M., Chamari, K., Amri, M., and Behm, D.G.Ten minutes of dynamic stretching is sufficient to potentiate vertical jump performance characteristics. *J Strength Cond Res* 25: 2453–2463, 2011.
142. Wallman, H.W., Mercer, J.A., and McWhorter, J.W.Surface Electromyographic Assessment of the Effect of static stretching of the gastrocnemius on vertical jump performance. *Journal of Strength and Conditioning Research* 19: 684–688, 2005.
143. Wallmann, H.W., Mercer, J.A., and Landers, M.R.Surface electromyographic assessment of the effect of dynamic activity and dynamic activity with static stretching of the gastrocnemius on vertical jump performance. *J Strength Cond Res* 22: 787–793, 2008.
144. Weiner, I.H. and Weiner, H.L.Nocturnal leg muscle cramps. *JAMA* 244: 2332–2333, 1980.
145. Wilson, G., Elliot, B., and Wood, G.Stretching shorten cycle performance enhancement through flexibility training. *Medicine and Science in Sports and Exercise* 24: 116–123, 1992.
146. Winchester, J.B., Nelson, A.G., Landin, D., Young, M.A., and Schexnayder, I.C. Static stretching impairs sprint performance in collegiate track and field athletes. *J Strength Cond Res* 22: 13–19, 2008.
147. Worrell, T., Smith, T., and Winegardner, J.Effect of hamstring stretching on hamstring muscle performance. *Journal of Orthopedic Sports Physical Therapy* 20: 154–159, 1994.
148. Yamaguchi, T. and Ishii, K.An optimal protocol for dynamic stretching to improve explosive performance. *Journal of Physical Fitness and Sports Medicine* 3, 2014.
149. Young, W. and Behm, D.Should static stretching be used during a warm-up for strength and power activities? *Strength and Conditioning Journal* 24: 33–37, 2002.
150. Young, W., Elias, G., and Power, J.Effects of static stretching volume and intensity on plantar flexor explosive force production and range of motion. *Journal of Sports Medicine and Physical Fitness* 46: 403–411, 2006.
151. Young, W. and Elliott, S.Acute effects on static stretching, proprioceptive neuromuscular facilitation stretching, and maximum voluntary contractions on explosive force production and jumping performance. *Research Quarterly for Exercise and Sport* 72: 273–279, 2001.
152. Young, W.B. and Behm, D.G.Effects of running, static stretching and practice jumps on explosive force production and jumping performance. *J Sports Med Phys Fitness* 43: 21–27, 2003.
153. Yuktasir, B. and Kaya, F.Investigation into the long term effects of static and PNF stretching exercises on range of motion and jump performance. *Journal of Bodywork and Movement Therapies* 13: 11–21, 2009.

10 Effect of Stretch Training on Functional Performance

Static Stretching

While acute prolonged static stretching can lead to impairments, there are physiological rationales why chronic static stretching could improve performance. Static stretch training can enhance Ca+ within the neuromuscular junction (37), decrease muscle stiffness (35), and proliferate sarcomeres in series. (9, 39). Increased neuromuscular junction Ca+ should enhance neurotransmitter release. Decreased musculotendinous stiffness or increased compliance could reduce the resistance to movement improving movement and energy efficiency. Energy efficiency would be directly related to a training-related decrease in hysteresis (18). Hysteresis is related to tissue viscoelastic changes affecting heat loss with stretching (14). Thus training-induced changes in viscoelasticity leading to decreased hysteresis (the return to the original state is delayed) would help retain rather than dissipate energy through heat loss (18). Furthermore, increased musculotendinous unit compliance could positively affect the stretch-shortening cycle (SSC) by enhancing the ability to store elastic energy (18, 35). This compliance-induced elastic storage enhancement would be more apparent with longer duration eccentric to concentric transitions such as a rebound chest press (35) but might not have such a positive impact on very short SSC activities like sprinting where the athlete might transition from foot strike to take-off in less than 100 ms.

A higher number of sarcomeres in series could allow muscle to produce more force or torque at longer muscle lengths and change the optimal force angle (8). It is reported that many of the musculotendinous injuries occur at extended muscle lengths when the muscle exerts lower forces and cannot protect itself as well against environmental stressors to as great an extent (4). Another advantage is that the number of sarcomeres in series is related to muscle contraction velocity (15). As a reminder though, increases in sarcomere number with stretching has been primarily verified in animal studies but not human in-vivo studies.

Although these proposed physiological adaptations should provide physical benefits, does the practical research concur? A well cited review by Shrier (24) reported that regular stretching would improve most athletic performances. Another systematic review (18) showed that only half of the stretch training

DOI: 10.4324/9781032709086-11

articles (14/28) improved muscle performance. There was very little effect upon static or isometric contraction measures. Muscle performance improvements in these studies were generally dynamic activities such as jumps, isokinetic eccentric and concentric torque, rebound bench press, and plantar flexors 1RM. However, even with dynamic activities the results were not consistent. Similar to the Behm et al. (4) review that suggested a major advantage of stretching would be enhanced force outputs at longer muscle lengths, the Medeiros and Lima review (18) found a number of studies reporting enhanced eccentric peak torque. As eccentric contractions are ubiquitous in all activities but especially important in SSC activities, as they help to potentiate or augment the concentric contraction (6, 7), a flexibility training-induced improvement in eccentric strength could be of vital importance. As you should suspect by now, none of these statements ever get unanimous support.

For example, while it is safe to state that almost every chronic stretch training programme increases range of motion (ROM) (13), there are a number of studies that reported no change in evoked contraction speed (8 weeks) (1), sprint time and vertical jump height (4–6 weeks) (3, 11, 22), countermovement jump height (nine weeks in pre-adolescent athletes) (10), drop jump height (four and six weeks) (19, 38), dynamic knee flexion strength (six weeks) (5), plantar flexors 1RM (ten weeks) (20), or maximum voluntary isometric contraction force (four, six, and 12 weeks) (17, 19, 22, 23). Some chronic stretch training studies actually show deficits, with for example decreases in eccentric peak torque and triple hop distance (ten sessions) (2). Just to increase the confusion, there are a number of studies demonstrating stretch training-induced improvements in hamstrings peak torque and countermovement jump height (eight weeks) (16), vertical jump height (four weeks: squat and countermovement jump) (21), dynamic postural stability (six weeks) (22) and MVIC force (five weeks) (36). Lima et al. (16) reported greater effect sizes (larger magnitude improvements) with an 8 week periodized stretch training programme versus non-periodized for ROM and performance. Why is there such diversity in these results? There was very little consistency in the stretching prescriptions in these studies. Prescriptions ranged from session durations of 2x30 seconds, 3x30s, 5x30s, 3x60s, 5x45s, or 2 minutes, weekly frequencies of one, three, four, or five days/week and training programme durations of ten sessions, 15 days, four, six, eight, ten, or 12 weeks. In these studies, we cannot see a trend for improved performance with longer training durations or higher weekly frequencies or even longer session durations. However, compared to some recent research showing stretch-induced increases in muscle strength and hypertrophy (26, 27, 28, 29, 30, 31, 32, 33, 34) these cited session durations are relatively short (<4 minutes) and often only three days per week. A recent review of chronic stretching effects on jump and sprint performance demonstrated a significant but trivial magnitude increase in jump height but no significant effect on sprint time or speed. We summarized and theorized that static stretching was not a sufficient stimulus to substantially enhance jump and sprint performance, possibly owing to small weekly training volumes or lack of intensity.

Table 10.1 Dynamic Stretching FITT table

FITT Recommendations	
Frequency	• Daily (≥5 per week provide optimal benefits, but improvements are still seen at the rate of twice per week)
Intensity	• Full range of motion (dynamic stretching: submaximal limb velocity)
Time	• 3–6 sets of 30–90 seconds with 15s rest periods between sets
Type	• Dynamic stretching for all major muscle-tendon units is recommended • Dynamic and Ballistic (higher-velocity movements typically with bouncing actions at the end ROM) flexibility

The next chapter will highlight the more consistently positive results on muscle strength gains when more prolonged static stretching durations (10–60 minutes per muscle group) are implemented into nearly daily (e.g., 5–6 days per week) stretch training programmes.

Dynamic Stretch Training

When dynamic stretching was incorporated into the daily warm-up of wrestlers for 4 weeks there were improvements in power, strength, muscular endurance, anaerobic capacity, and agility performance (12). Similarly, an eight-week programme involving either active (not staying in one spot) or static (staying in one location) dynamic stretch training improved both flexibility and jump parameters but not sprint performance (25). However, there are also reports of no changes of eccentric peak torque and triple hop test distance (3x week for ten sessions) (2).

Summary

In contrast to the reported performance impairments with acute bouts of prolonged static stretching, there are many physiological adaptations with static stretch training that should enhance performance. However, the evidence from these low volume or duration static stretch training studies is contradictory. The few dynamic stretch studies do show performance improvements but not with every measure.

References

1. Alvarez-Yates, T. and Garcia-Garcia, O.Effect of a Hamstring Flexibility Program Performed Concurrently During an Elite Canoeist Competition Season. *J Strength Cond Res* 34: 838–846, 2020.
2. Barbosa, G.M., Trajano, G.S., Dantas, G.A.F., Silva, B.R., and Vieira, W.H.B. Chronic Effects of Static and Dynamic Stretching on Hamstrings Eccentric Strength

and Functional Performance: A Randomized Controlled Trial. *J Strength Cond Res* 34: 2031–2039, 2020.

3. Bazett-Jones, D.M., Gibson, M.H., and McBride, J.M.Sprint and vertical jump performances are not affected by six weeks of static hamstring stretching. *J Strength Cond Res* 22: 25–31, 2008.
4. Behm, D.G., Blazevich, A.J., Kay, A.D., and McHugh, M.Acute effects of muscle stretching on physical performance, range of motion, and injury incidence in healthy active individuals: a systematic review. *Appl Physiol Nutr Metab* 41: 1–11, 2016.
5. Brusco, C.M., Blazevich, A.J., and Pinto, R.S.The effects of 6 weeks of constant-angle muscle stretching training on flexibility and muscle function in men with limited hamstrings' flexibility. *Eur J Appl Physiol* 119: 1691–1700, 2019.
6. Cappa, D.F. and Behm, D.G.Training Specificity of Hurdle vs. Countermovement Jump Training. *Journal of Strength and Conditioning Research/National Strength & Conditioning Association*, 2011.
7. Cappa, D.F. and Behm, D.G.Neuromuscular characteristics of drop and hurdle jumps with different types of landings. *J Strength Cond Res* 27: 3011–3020, 2013.
8. Chen, T.C., Lin, K.Y., Chen, H.L., Lin, M.J., and Nosaka, K.Comparison in eccentric exercise-induced muscle damage among four limb muscles. *Eur J Appl Physiol* 111: 211–223, 2011.
9. De Deyne, P.G.Application of passive stretch and its implications for muscle fibers. *Phys Ther* 81: 819–827, 2001.
10. Donti, O., Papia, K., Toubekis, A., Donti, A., Sands, W.A., and Bogdanis, G.C. Acute and long term effects of two different static stretching training protocols on range of motion and vertical jump in preadolescent athletes. *Biology of Sport* 38: 579–586, 2021.
11. Gunaydin, G., Citaker, S., and Cobanoglu, G.Effects of different stretching exercises on hamstring flexibility and performance in long term. *Science and Sports*, 2020.
12. Herman, S.L. and Smith, D.T.Four-week dynamic stretching warm-up intervention elicits longer-term performance benefits. *J Strength Cond Res* 22: 1286–1297, 2008.
13. Konrad, A., Alizadeh, S., Daneshjoo, A., Anvar, S.H., Graham, A., Zahiri, A., Goudini, R., Edwards, C., Scharf, C., and Behm, D.G.Chronic effects of stretching on range of motion with consideration of potential moderating variables: A systematic review with meta-analysis. *J Sport Health Sci*, 2023.
14. Kubo, K.In Vivo Elastic Properties of Human Tendon Structures in Lower Limb. *International Journal of Sport and Health Science* 3: 143–151, 2005.
15. Lieber, R.L. and Bodine-Fowler, S.C.Skeletal muscle mechanics: implications for rehabilitation. *Phys Ther* 73: 844–856, 1993.
16. Lima, C.D., Brown, L.E., Li, Y., Herat, N., and Behm, D.Periodized versus Non-periodized Stretch Training on Gymnasts Flexibility and Performance. *Int J Sports Med* 40: 779–788, 2019.
17. Longo, S., Ce, E., Bisconti, A.V., Rampichini, S., Doria, C., Borrelli, M., Limonta, E., Coratella, G., and Esposito, F.The effects of 12 weeks of static stretch training on the functional, mechanical, and architectural characteristics of the triceps surae muscle-tendon complex. *Eur J Appl Physiol* 121: 1743–1758, 2021.
18. Medeiros, D.M. and Lima, C.S.Influence of chronic stretching on muscle performance: Systematic review. *Hum Mov Sci* 54: 220–229, 2017.
19. Nakamura, M., Yoshida, R., Sato, S., Yahata, K., Murakami, Y., Kasahara, K., Fukaya, T., Takeuchi, K., Nunes, J.P., and Konrad, A.Comparison Between High-

and Low-Intensity Static Stretching Training Program on Active and Passive Properties of Plantar Flexors. *Front Physiol* 12: 796497, 2021.

20. Nelson, A.G., Kokkonen, J., Winchester, J.B., Kalani, W., Peterson, K., Kenly, M. S., and Arnall, D.A.A 10-week stretching program increases strength in the contralateral muscle. *J Strength Cond Res* 26: 832–836, 2012.
21. Sakai, S., Maeda, N., Sasadai, J., Kotoshiba, S., Anami, K., Tashiro, T., Fujishita, H., and Urabe, Y.Effect of 4-week cyclic stretching program on muscle properties and physical performance in healthy adult men. *J Sports Med Phys Fitness* 60: 37–44, 2020.
22. Sasadai, J., Maeda, N., Shogo, S., Tsubasa, T., Hitoshi, A., and Yukio, U.Effect of a 4 week static stretching program for plantar flexor muscles on physical performance and muscle properties. *Isokinetics and Exercise Science* 23: 1–8, 2021.
23. Sato, S., Hiraizumi, K., Kiyono, R., Fukaya, T., Nishishita, S., Nunes, J.P., and Nakamura, M.The effects of static stretching programs on muscle strength and muscle architecture of the medial gastrocnemius. *PLoS One* 15: e0235679, 2020.
24. Shrier, I.Does stretching improve performance?: a systematic and critical review of the literature. *Clin J Sport Med* 14: 267–273, 2004.
25. Turki-Belkhiria, L., Chaouachi, A., Turki, O., Chtourou, H., Chtara, M., Chamari, K., Amri, M., and Behm, D.G.Eight weeks of dynamic stretching during warm-ups improves jump power but not repeated or single sprint performance. *Eur J Sport Sci* 14: 19–27, 2014.
26. Warneke, K., Brinkmann, A., Hillebrecht, M., and Schiemann, S.Influence of Long-Lasting Static Stretching on Maximal Strength, Muscle Thickness and Flexibility. *Front Physiol* 13: 878955, 2022.
27. Warneke, K., Keiner, M., Hillebrecht, M., and Schiemann, S.Influence of One Hour versus Two Hours of Daily Static Stretching for Six Weeks Using a Calf-Muscle-Stretching Orthosis on Maximal Strength. *Int J Environ Res Public Health* 19, 2022.
28. Warneke, K., Keiner, M., Wohlann, T., Lohmann, L.H., Schmitt, T., Hillebrecht, M., Brinkmann, A., Hein, A., Wirth, K., and Schiemann, S.Influence of Long-Lasting Static Stretching Intervention on Functional and Morphological Parameters in the Plantar Flexors: A Randomized Controlled Trial. *J Strength Cond Res* 37: 1993–2001, 2023.
29. Warneke, K., Konrad, A., Keiner, M., Zech, A., Nakamura, M., Hillebrecht, M., and Behm, D.G.Using Daily Stretching to Counteract Performance Decreases as a Result of Reduced Physical Activity-A Controlled Trial. *Int J Environ Res Public Health* 19, 2022.
30. Warneke, K., Lohmann, L.H., Keiner, M., Wagner, C.M., Schmidt, T., Wirth, K., Zech, A., Schiemann, S., and Behm, D.Using Long-Duration Static Stretch Training to Counteract Strength and Flexibility Deficits in Moderately Trained Participants. *Int J Environ Res Public Health* 19, 2022.
31. Warneke, K., Lohmann, L.H., Lima, C.D., Hollander, K., Konrad, A., Zech, A., Nakamura, M., Wirth, K., Keiner, M., and Behm, D.G.Physiology of Stretch-Mediated Hypertrophy and Strength Increases: A Narrative Review. *Sports Med*, 2023.
32. Warneke, K., Wirth, K., Keiner, M., Lohmann, L.H., Hillebrecht, M., Brinkmann, A., Wohlann, T., and Schiemann, S.Comparison of the effects of long-lasting static stretching and hypertrophy training on maximal strength, muscle thickness and flexibility in the plantar flexors. *Eur J Appl Physiol* 123: 1773–1787, 2023.
33. Warneke, K., Zech, A., Wagner, C.M., Konrad, A., Nakamura, M., Keiner, M., Schoenfeld, B.J., and Behm, D.G.Sex differences in stretch-induced hypertrophy, maximal strength and flexibility gains. *Front Physiol* 13: 1078301, 2022.

34. Warneke, K., Hillebrecht, M., Claasen-Helmers, E., Wohlann, T., Keiner, M., and Behm, D.G.Effects of a Home-Based Stretching Program on Bench Press Maximum Strength and Shoulder Flexibility. *Journal of Sport Science and Medicine* 22: 597–604, 2023.
35. Wilson, G., Elliot, B., and Wood, G.Stretching shorten cycle performance enhancement through flexibility training. *Medicine and Science in Sports and Exercise* 24: 116–123, 1992.
36. Yahata, K., Konrad, A., Sato, S., Kiyono, R., Yoshida, R., Fukaya, T., Nunes, J.P., and Nakamura, M.Effects of a high-volume static stretching programme on plantar-flexor muscle strength and architecture. *Eur J Appl Physiol* 121: 1159–1166, 2021.
37. Yamashita, T., Ishii, S., and Oota, I.Effect of muscle stretching on the activity of neuromuscular transmission. *Med Sci Sports Exerc* 24: 80–84, 1992.
38. Yuktasir, B. and Kaya, F.Investigation into the long term effects of static and PNF stretching exercises on range of motion and jump performance. *Journal of Bodywork and Movement Therapies* 13: 11–21, 2009.
39. Zollner, A.M., Abilez, O.J., Bol, M., and Kuhl, E.Stretching skeletal muscle: chronic muscle lengthening through sarcomerogenesis. *PLoS One* 7: e45661, 2012.

11 Effects of Stretch Training on Muscle Strength and Hypertrophy

Konstantin Warneke

Chronic stretch training is considered one of the most common training interventions to enhance the range of motion (ROM). A plethora of research has been conducted over the years to investigate the effects of different stretching interventions, concluding that static stretching was one of the most effective techniques. While previous chapters provided definitions of different stretching methods that are used in human research, a more physiological point of view might "stretch" our perspective of this research field. When opening a new scientific field of potential interest, it is very common to start investigations using animal models (7). Before diving into a recently emerged field of human research, we will introduce the history of stretching research by briefly providing early evidence from studies with birds, cats, and rabbits.

What humans can learn from chicken and quails

The foundation of probably most of the known stretching research was established as early as 1887, using animal models. Marey (10) removed the attachment of the soleus muscle and attached it farther away from the skeletal origin. While this surgery might not be considered a relevant training intervention, the muscle was chronically stretched and responded with a rapid enhancement in the number of sarcomeres in series, (sarcomerogenesis), which resulted in an enhanced muscle length. However, to induce a chronic mechanical overload, it is not necessary to invasively move the muscle attachment. I am sure you can imagine what the human ethics committees might say if we tried these experiments on our students or the public. A series of investigations starting in the second half of the 20th century applied chronic stretches to the wing muscles of chickens and quails by attaching weights or stretching apparatus that abducted the adductors. By adopting very different protocols, using stretching durations that ranged from 30 minutes per day for five weeks to chronic interventions of 24 hours per day, seven days per week for up to six weeks, the various researchers consistently showed large muscle mass increases in the stretched muscle (18, 19, 20, 21, 22), which followed a duration-response relationship (more stretching = more hypertrophy). However, the enhanced muscle mass was not exclusively attributable to

DOI: 10.4324/9781032709086-12

serial sarcomere attachments, or in other words, increases in muscle length. As early as after the first day of stretching overload, the authors reported additional muscle and fibre cross-sectional increases, while chronic training protocols resulted in a significantly increased number of muscle fibres (hyperplasia). These increases were accompanied by a significant increase in muscle protein DNA and RNA resulting in a stimulated net protein synthesis rate, resulting in an increase in total amount of protein, RNA and DNA by 59 percent, 228 percent, and 82 percent respectively after seven days (17) that were attributed to mechanical tissue overload from passive stretching. However, these results were questioned, as it could not be ruled out that attaching a weight or stretching apparatus to the birds would initiate active muscle contractions. To explore the role of active contractions, Sola, Christensen, and Martin (16) removed the innervating nerve for the stretched muscle and compared the adaptations with those of innervated wing muscles. In both groups, there were substantial muscle mass increases with a slightly smaller effect in the denervated muscle. Accordingly, in 1980, Barnett and colleagues (1) and Holly et al. (6) confirmed no electromyography increases in the chronically stretched muscle, underlining the potential of stretch-induced mechanical overload. Thus, even prolonged passive stretching with its tension on the tissues that are typically lower than experienced with resistance training can induce muscle hypertrophy. For more detailed underlying mechanisms and the results of individual studies can be reviewed in Warneke et al. (23). Briefly, the tension on muscle whether it is from resistance training or static stretching will trigger the integrin molecules on the muscle membranes to initiate a cascade of events that initiate protein synthesis through the Raptor/mTOR complex. I find an easy way to remember the function of the integrin molecule is to remember that it monitors the "INTEGRITY" of the membrane and thus if the integrity is being threatened by tension, it triggers protein synthesis to protect it. While active contractions will add a number of other contributing mechanisms, passive tension with stretching alone can still promote muscle hypertrophy.

Interestingly, Frankeny et al. (5) and Bates (2) showed that chronic (24 hours/7 days per week) stretching interventions in animal models were unnecessary when increasing muscle strength and mass. Both studies used intermittent stretching protocols and confirmed significant muscle mass increases after at least 30 minutes of stretch. Nonetheless, longer stretching durations might be not economical. Bates (2) allocated 30 chickens into different stretching groups that performed daily anterior latissimus dorsi stretching for five weeks. While 30 minutes daily stretching resulted in 57 percent mass increase, doubling the stretching time provided only additional 2 percent. Another doubling of the stretching time (two hours of daily stretching) resulted in a 67 percent increase, while an additional two hours showed 72 percent muscle mass increase after four hours of daily stretching. Also Frankeny, Holly and Ashmore (5: pp. 275–276) summarized that "Thirty minutes of stretching per day is certainly within normal physiological limits, and as a result may be applied to human muscle with hopes that similar adaptations would occur". The authors suggested exploring the transferability of animal results to humans. Two reasons might explain the lack of longer duration stretch training studies in humans until 2021. The first reason associated with animal data collection procedures is obvious. All animal participants are terminated after the prescribed intervention period, with the muscle removed and assessed via light microscopy. Indubitably, recruiting human volunteers for such a study might be a little difficult! ☺ Secondly, it is difficult to recruit enough participants who are willing to spend up to several hours per day stretching, in addition to their other activities of daily life.

Does stretching provide a sufficient stimulus to enhance human muscle strength and size?

These experimental burdens prevented designing studies that explored the transferability of these animal stretch results to humans. While an early review article suggested a possible benefit of stretch training for performance parameters such as jumping, sprinting, or strength-related performance (14), human research was rare. Kokkonen and colleagues (9) stretched lower extremity muscles with 3x15 seconds per muscle for an overall training time of 40 minutes per session, three days per week for ten weeks and demonstrated significant increases of 32.4 percent in the knee extensors, 15.3 percent in the knee flexors, with improvements in sprinting and jumping performance. Recruiting female track and field athletes, Bazett-Jones et al. (3) did not confirm the benefits of stretching on athletic performance. Within the following years, other researchers conducted a variety of static stretching interventions on muscle strength and hypertrophy, resulting in inconclusive results with studies showing stretch-induced strength and muscle size improvements, while others did not. These controversial results might arise from a high heterogeneity in the stretching methods that prevented Medeiros and Lima (11) from performing a meta-analysis. While in 2017 Medeiros and Lima (11) reported a potential benefit of stretching on performance, in 2020, Nunes et al. (12) reviewed the existing literature without confirming stretch-mediated hypertrophy in humans. However, the authors highlighted that a lack of research performed high intensity/high volume stretching interventions. Indeed, it would be inconsistent to compare five minutes or less of continuous stretching per session in humans, with animal research that implemented a minimum of 30 minutes daily stretching.

Interestingly, in 2021 researchers from Japan and Greece initiated stretching studies using a stretching board for 6x5 minutes, two days per week or up to 5x15 minutes for five days per week (13). While in the Yahata (27) study, strength increases of about 10 percent were not accompanied by significant muscle hypertrophy, Panidi and colleagues (13) reported large effect size magnitudes (24 percent) of muscle hypertrophy. Both studies suggested the applied stretching dosage to be of crucial importance for stretching responses regarding strength and hypertrophy.

In 2022 the first studies using training loads more similar to animal research were conducted (18, 19, 20, 21, 22). Applying stretching partly via a stretching orthosis, these works performed stretching for 5–120 minutes per session, 6–7 days per week for six weeks. By using ultrasound- and magnetic resonance imaging (MRI), significant stretch-induced muscle hypertrophy in the plantar flexors (16 percent), that were accompanied by maximal strength increases of up to 25 percent were confirmed.

Underlying mechanisms of stretch-mediated hypertrophy and strength increases

In contrast to animal research that showed an improved muscle protein synthesis that was attributed to mechanical overload-induced stimulation of anabolic signalling pathways (23), the underlying physiology in humans is rarely investigated. In 1993 a study by Smith and colleagues (15) described potential indications of mechanical overuse induced with stretching by finding enhanced creatine kinase values. However, these results could not be confirmed by Wohlann et al. (25, 26) Moreover, the only stretching study that assessed stretch-induced protein synthesis changes was not able to find significant changes (4). However, as these studies did not use appropriate stretching volumes (e.g., insufficient durations with an overall volume of 33 minutes), the role of stretching-induced mechanical overload and subsequent enhanced protein synthesis remains speculative in humans.

Pointing in another direction, Hotta et al. (8) stretched rat hindlimbs for 30 minutes per day for four weeks and found blood flow restrictions while stretching, which caused an enhanced capillarization. Also in humans, the effects of restricted blood flow on muscle hypertrophy are well investigated. Even though the literature lacks evidence that reports decreased blood flow while performing 30–120 minutes of stretching per day in humans, it seems a concurrent, but logical alternative explanation.

Consequently, current systematic reviews report significant, small magnitude improvements in maximal strength and muscle size with static stretching, if performed with sufficient volume, intensity, and frequency. However, further research is required to understand the underlying physiological processes to maximize and optimize these effects.

Practical applications and limitations

Although it is possible to enhance muscle strength and size without performing active contractions to a similar magnitude as with commonly performed resistance training in the plantar flexors and pectoralis muscle, one might argue that there is limited practical relevance, owing to the comparatively long durations required with static stretching. To reach comparable results in the calf muscles, seven hours of stretching, four days per week were necessary, while Wohlann and colleagues (25, 26) used a stretching apparatus and performed 15 minutes of uncomfortable stretching including constant re-adjustment (increasing) of the stretching tension to counteract creeping and relaxation effects. Especially as these training volumes were applied to just one muscle, the exercise routine seems of limited economy (time restrictions) compared with active resistance training, that may use 20 percent of stretch training durations to reach almost similar results. Nevertheless, there might be some populations with restricted accessibility to training equipment, or after serious injuries that cause immobilization of a limb, for whom active resistance training might be contra-indicated. For those individuals, but also the unmotivated populations, high volume of passive stretching by using a stretching device that enables the integration of this workout into the daily seated times while watching television or working in the office could provide a valuable supplementation. Active movements however, if applicable might provide further benefits such as improved cardiovascular health, neuromuscular adaptations, or transfer effects to athletic performance parameters such as sprinting or jumping. Even though some high-volume stretching studies showed first indications of enhanced squat jump performance by stretching, Warneke et al. (24) were not able to replicate the benefits of chronic stretching routines on jumping or sprinting performance. Thus, using active exercise routines would be preferred when aiming to enhance strength and hypertrophy, when applicable, but there are scenarios where stretch-induced muscle strength and hypertrophy may be a suitable alternative.

References

1. Barnett, J.G., Holly, R.G., and Ashmore, C.R. (1980). Stretch-induced growth in chicken wing muscles: biochemical and morphological characterization. *Am J Physiol* 239: C39–46.
2. Bates, G.P. (1993). The relationship between duration of stimulus per day and the extent of hypertrophy of slow-tonic skeletal muscle in the fowl, *Gallus gallus. Comp Biochem Physiol Comp Physiol* 106: 755–758.
3. Bazett-Jones, D.M., Gibson, M.H., and McBride, J.M. (2008). Sprint and vertical jump performances are not affected by six weeks of static hamstring stretching . *J Strength Cond Res* 22: 25–31.
4. Fowles, J.R., MacDougall, J.D., Tarnopolsky, M.A., Sale, D.G., Roy, B.D., and Yarasheski, K.E. (2000). The effects of acute passive stretch on muscle protein synthesis in humans. *Canadian Journal of Applied Physiology* 25: 165–180.
5. Frankeny, J.R., Holly, R.G., and Ashmore, C.R. (1983). Effects of graded duration of stretch on normal and dystrophic skeletal muscle. *Muscle Nerve* 6: 269–277.
6. Holly, R.G., Barnett, J.G., Ashmore, C.R., Taylor, R.G., and Mole, P.A. (1980). Stretch-induced growth in chicken wing muscles: a new model of stretch hypertrophy. *Am J Physiol* 238: C62–71.
7. Hooijmans, C.R., IntHout, J., Ritskes-Hoitinga, M., and Rovers, M.M. (2014). Meta-analyses of animal studies: an introduction of a valuable instrument to further improve healthcare. *ILAR J* 55: 418–426.
8. Hotta, K., Behnke, B.J., Arjmandi, B., Ghosh, P., Chen, B., Brooks, R., Maraj, J.J., Elam, M.L., Maher, P., Kurien, D., Churchill, A., Sepulveda, J.L., Kabolowsky, M.

B., Christou, D.D., and Muller-Delp, J.M. (2018). Daily muscle stretching enhances blood flow, endothelial function, capillarity, vascular volume and connectivity in aged skeletal muscle. *J Physiol* 596: 1903–1917.

9. Kokkonen, J., Nelson, A.G., Eldredge, C., and Winchester, J.B. (2007). Chronic static stretching improves exercise performance. *Med.Sci.Sports Exerc.* 39: 1825–1831.
10. Marey, E. (1887). Recherches experimentales sur la morphologie des muscles. *CR Academy of Sciences* 105: 445–451.
11. Medeiros, D.M. and Lima, C.S. (2017). Influence of chronic stretching on muscle performance: Systematic review. *Hum Mov Sci* 54: 220–229.
12. Nunes, J.P., Schonefeld, B.J., Nakamura, M., Ribiero, A.S., Cunha, P.M., and Cyrino, E.S. (2020). Does stretch training induce muscle hypertrophy in humans? A review of the literature? *Clincal Physiology andd Functional Imaging* 40: 148–156.
13. Panidi, I., Bogdanis, G.C., Terzis, G., Donti, A., Konrad, A., Gaspari, V., and Donti, O. (2021). Muscle Architectural and Functional Adaptations Following 12-Weeks of Stretching in Adolescent Female Athletes. *Front Physiol* 12: 701338.
14. Shrier, I. (2004). Does stretching improve performance?: a systematic and critical review of the literature. *Clin.J.Sport Med.* 14: 267–273.
15. Smith, L.L., Brunetz, M.H., Chenier, T.C., McCammon, M.R., Houmard, J.A., Franklin, M.E., and Israel, R.G. (1993). The effects of static and ballistic stretching on delayed onset muscle soreness and creatine kinase. *Research Quarterly for Exercise and Sport* 64: 103–107.
16. Sola, O.M., Christensen, D.L., and Martin, A.W. (1973). Hypertrophy and hyperplasia of adult chicken anterior latissimus dorsi muscles following stretch with and without denervation. *Exp Neurol* 41: 76–100.
17. Sparrow, M. (1982). Regression of skeletal muscle of chicken wing after stretch-induced hypertrophy. *American Journal of Physiology* 242: C333–C338.
18. Warneke, K., Brinkmann, A., Hillebrecht, M., and Schiemann, S. (2022a). Influence of Long-Lasting Static Stretching on Maximal Strength, Muscle Thickness and Flexibility. *Front Physiol* 13: 878955.
19. Warneke, K., Keiner, M., Hillebrecht, M., and Schiemann, S. (2022b). Influence of One Hour versus Two Hours of Daily Static Stretching for Six Weeks Using a Calf-Muscle-Stretching Orthosis on Maximal Strength. *Int J Environ Res Public Health* 19.
20. Warneke, K., Keiner, M., Wohlann, T., Lohmann, L.H., Schmitt, T., Hillebrecht, M., Brinkmann, A., Hein, A., Wirth, K., and Schiemann, S. (2023a). Influence of Long-Lasting Static Stretching Intervention on Functional and Morphological Parameters in the Plantar Flexors: A Randomized Controlled Trial. *J Strength Cond Res* 37: 1993–2001.
21. Warneke, K., Konrad, A., Keiner, M., Zech, A., Nakamura, M., Hillebrecht, M., and Behm, D.G. (2022c). Using Daily Stretching to Counteract Performance Decreases as a Result of Reduced Physical Activity-A Controlled Trial. *Int J Environ Res Public Health* 19.
22. Warneke, K., Wirth, K., Keiner, M., Lohmann, L.H., Hillebrecht, M., Brinkmann, A., Wohlann, T., and Schiemann, S. (2023c). Comparison of the effects of long-lasting static stretching and hypertrophy training on maximal strength, muscle thickness and flexibility in the plantar flexors. *Eur J Appl Physiol* 123: 1773–1787.
23. Warneke, K., Lohmann, L.H., Lima, C.D., Hollander, K., Konrad, A., Zech, A., Nakamura, M., Wirth, K., Keiner, M., and Behm, D.G. (2023b). Physiology of Stretch-Mediated Hypertrophy and Strength Increases: A Narrative Review. *Sports Medicine* 53: 2055–2075.

24. Warneke, K., Freundorfer, P., Ploschberger, G., Konrad, A., Behm, D.G., Schmidt, T. (2024). Effects of chronic static stretching interventions on jumping and sprinting performance –- A Systematic Review with Multilevel Meta-Analysis. *Frontiers in Physiology* (in press forthcoming).
25. Wohlann, T., Warneke, K., Hillebrecht, M., Petersmann, A., Ferrauti, A., and Schiemann, S. (2023). Effects of daily static stretch training over 6 weeks on maximal strength, muscle thickness, contraction properties, and flexibility. *Front Sports Act Living* 5: 1139065.
26. Wohlann, T., Warneke, K., Kalder, V., Behm, D.G., Schmidt, T., and Schiemann, S. (2024). Influence of 8 weeks of supervised static stretching or resistance training of pectoral major muscles on maximal strength, muscle thickness and range of motion. *Eur J Appl Physiol.*
27. Yahata, K., Konrad, A., Sato, S., Kiyono, R., Yoshida, R., Fukaya, T., Nunes, J.P., and Nakamura, M. (2021). Effects of a high-volume static stretching programme on plantar-flexor muscle strength and architecture. *Eur J Appl Physiol* 121: 1159–1166.

Section II

The Science and Physiology of Alternative Techniques for Increasing Range of Motion

12 Effects of Resistance Training on Range of Motion

Shahab Alizadeh

The historical roots of strength training can be traced back to ancient civilizations, with the earliest recorded instances dating back to the Chou dynasty in China around 1122–249 BCE (10). However, a more structured approach to strength training emerged during the 6th century BCE with the legend of Milo of Croton, a Greek wrestler. Milo's remarkable athletic accomplishments, which included an impressive tally of 32 wrestling victories, were purportedly influenced by his unorthodox approach to building strength. According to legend, it was said that Milo would carry a young calf on his shoulders every day. As the calf grew into a fully grown and heavier animal over time, so too did Milo's strength increase proportionally. While some may question the historical accuracy surrounding this account, its significance lies in its symbolic representation of progressive resistance training – an essential concept revolving around systematically increasing one's workload with the aim of bolstering muscle strength, power, and endurance.

Milo's tale acts as an early precursor in history for our modern comprehension and appreciation of strength training methodologies centred around progressive overload principles. Despite being anecdotal in nature, this story underscores ancient insights recognizing the association between resistance-based exercises and their potential impact on augmenting overall muscular strength – ultimately providing a fundamental groundwork upon which further advancements within exercise science and physical conditioning could be built upon.

The origins of research on resistance training can be traced back to the late 1800s. For example, the work conducted by Scripture and colleagues in 1894 (29), regarding the cross-education effect of strength training represented an early exploration into a phenomenon where individuals experienced improvements not only in their trained limb but also in their untrained limb. This discovery suggested that strength adaptations were not limited to specific limbs but had systemic or at least contralateral implications, which served as a basis for future investigations into the broader effects of resistance training as well as more recently; global effects of stretching (see Chapter 6).

However, it was during the mid-1900s that resistance training research underwent its most significant paradigm shift. This period witnessed a surge of diverse studies focusing on different aspects of resistance training. Scholars

DOI: 10.4324/9781032709086-14

and scientists devoted themselves to comprehending and optimizing various elements (e.g., resistance training volume, intensity, frequency, recovery periods, movement velocity and others) associated with this type of exercise, to enhance strength, power output, and deciphering the physiological mechanisms underlying muscle hypertrophy.

However, even in the 1970s resistance training was viewed with scepticism as it was believed to lead to muscle-boundness. Muscle-boundness referred to more stiff, hypertrophied muscles, which impaired muscle elasticity (flexibility), speed, and coordination. Hence, resistance training was not recommended for most athletes. In 1964 J.R. Leighton's research shed light on an unexpected aspect of resistance training (19). While the primary focus of such training had traditionally been on augmenting physical strength, his study revealed that individuals engaging in resistance training not only experienced increases in strength but also demonstrated a significant enhancement in their range of motion (ROM). Leighton's work marked a paradigm shift, challenging the conventional view of resistance training as a unidimensional approach solely geared towards building muscular strength and size. The acknowledgment of improved ROM as an unintended yet consequential outcome broadened the scope of understanding surrounding the physiological adaptations associated with resistance training.

This revelation underscored the holistic impact of resistance training on human movement and functional capacity, emphasizing that its benefits extended beyond mere muscular development. By recognizing the interplay between strength gains and enhanced flexibility, Leighton's research contributed to a more nuanced appreciation of the multifaceted advantages associated with resistance training in the context of comprehensive physical fitness. This recognition, rooted in empirical evidence, laid the groundwork for subsequent scholarly inquiries into the diverse physiological adaptations resulting from resistance training.

Moderating Variables

When examining how strength training affects flexibility (ROM), it's important to distinguish between different types of resistance exercises. The primary isoinertial (resistance training that maintains a constant inertia throughout the ROM, facilitating a relatively constant resistance at every angle) types use free weights, machines, resistance bands, and body weight (e.g., calisthenics). Each of these methods has its own merits, but not all will improve the ROM of an individual (1).

Our meta-analysis showed that all types of strength training, except for using just your body weight (e.g., calisthenics such as push-ups), can increase ROM (1). When considering the mechanics of free weights, machines, and Pilates resistance training, one might suggest that these actions are similar to dynamic stretching, albeit with an additional external load (1). Dynamic stretching is an action that involves a controlled movement through the active joint ROM (3, 4, 9). Free weights, machines, and resistance bands resistance training typically allows the joints to reach their endpoint ROM or the individual's point of initial or maximum discomfort at a controlled pace. Remember, research shows that it is unnecessary to go to the maximum ROM or discomfort to achieve gains in ROM (2, 3, 6). Hence, it is also unnecessary to move heavy weights to an extreme ROM where according to the force-length relationship of muscle (8, 14, 21), it is at its weakest point and thus would be more susceptible to injury.

In contrast, resistance training with body mass activities may not always permit such an expansive ROM. For example, while a push-up is restricted by chest circumference and the surface (i.e., floor or ground), free weights and machines can permit the shoulder to surpass this more restricted ROM. There is a dearth of studies comparing the effects of partial versus full ROM resistance training on ROM. Kawama et al. (16) reported significantly greater decreases in the shear modulus (muscle stiffness) of the semimembranosus (hamstrings muscle) with eccentric resistance training through a wide ROM versus either eccentric resistance training through a narrow ROM or concentric resistance training with a wide ROM. Furthermore, dynamic stretching involves repeated cyclical muscle loading and unloading (9). The addition of an external load with RT would augment the stress on musculotendinous and connective tissue and minimize the unloading component. As there are no studies comparing dynamic stretch training and resistance training on ROM, possible differential effects of these two activities on ROM should be investigated.

When examining additional factors that may impact improvements in ROM through resistance training, the training status of individuals emerges as a significant moderating variable. Notably, there is a discernible influence on outcomes based on whether individuals are categorized as sedentary or active. Surprisingly, sedentary individuals have exhibited greater improvements in ROM with resistance training compared to their active counterparts (1). This discrepancy can be linked to the baseline level of flexibility, which sets the stage for the greater magnitude of training-induced ROM enhancements. The observed trend can be explained by considering the pre-existing conditions of musculotendinous units. Sedentary individuals, starting with a lower baseline of flexibility, appear to experience more substantial gains in ROM as a result of resistance training. In contrast, individuals with an active lifestyle, particularly those engaged in regular dynamic loading exercises, may already have an elevated baseline of flexibility. Consequently, their potential for training-induced ROM increases could be somewhat muted compared to those who were previously untrained (20). It is noteworthy, however, that even individuals with a trained background still undergo significant improvements in ROM, albeit to a lesser degree than their untrained counterparts. This emphasizes the adaptability of the human body and the continued potential for enhanced flexibility through targeted resistance training, even for those who have a history of regular physical activity. Understanding how training status interacts with the outcomes of resistance training on ROM is pivotal for tailoring effective training programmes based on individual characteristics and goals.

It is surprising that factors such as training frequency, training duration, age, and sex do not appear to significantly influence the outcomes of improvements in ROM through resistance training (1). Furthermore, it has been observed that the magnitude of ROM improvements does not exhibit notable variations across different joints (1). This finding implies that the benefits of resistance training for ROM are relatively consistent across various joints. Unlike some fitness outcomes that might show joint-specific responses, the impact of resistance training on flexibility appears to be more uniformly distributed.

Understanding these moderating variables provides valuable insights for designing comprehensive resistance training programmes aimed at improving flexibility. The lack of pronounced effects based on training frequency and sex suggests that individuals, regardless of their sex or how often they engage in resistance training, can experience comparable gains in ROM. Moreover, the uniformity in ROM improvements across joints underscores the holistic nature of resistance training's impact on flexibility throughout the body.

Possible Mechanisms

Various theories have been proposed to explain why resistance training improves ROM. In the context of neural adaptations associated with stretching and resistance training, particularly isoinertial resistance training, researchers have explored various mechanisms. Reports suggest that static stretch training over three to six weeks reduces tonic Ia afferent feedback from muscle spindles, which affects T-reflexes and H-reflexes (7, 11). This reduction in afferent feedback could potentially lead to diminished reflex-induced contractions, resulting in a more relaxed muscle state, referred to as disfacilitation. However, it's worth noting that dynamic stretching and isoinertial resistance training are more likely to excite, rather than disfacilitate, muscle spindle activity. While Golgi tendon organ inhibition is plausible with large amplitude stretches and higher muscle tension (12), it tends to subside immediately after the stimulus discontinues, making it an unlikely candidate for chronic dynamic stretching or isoinertial resistance training mechanisms.

Recurrent or Renshaw cell inhibition, which is associated more with acute dynamic rather than tonic contractions (15), may induce stabilizing effects on motoneuron discharge variability and motor unit synchronization (27). However, it's unclear whether these acute neural responses lead to chronic training adaptations.

Morphologically, dynamic ballistic stretch training has shown some evidence of decreasing muscle and tendon stiffness (25). Acute dynamic stretching has been associated with decreased passive resistive torque (17) and muscle stiffness (13, 17), suggesting a more compliant musculotendinous unit following a single session. However, a six-week ballistic stretch training programme did not reveal significant changes in muscle morphology (18).

Furthermore, there's a debate regarding the effects of resistance training on tendon and muscle stiffness. Some suggest that resistance training increases tendon stiffness through modifications in muscle elasticity (30), while others argue for an increase in muscle stiffness with resistance training (22). Eccentric actions with resistance training, which involve force production at extended positions, may increase ROM by altering fascicle length (26, 28) and pennation angle (28).

Notably, there is evidence supporting the increase in stretch (pain) tolerance with stretching (5, 23, 24) and a similar increase in pain tolerance is speculated with isoinertial resistance training, owing to the discomfort associated with external torques on muscles and joints. The combined impact of changes in musculotendinous unit stiffness and compliance, altered stretch tolerance, and potential adjustments in fascicle length and pennation angle may collectively contribute to the observed increase in ROM with dynamic stretching under load, such as isoinertial resistance training.

Summary

The evidence suggests that resistance training with external loads alone has the potential to improve ROM, making additional stretching either before or after resistance training potentially unnecessary for enhancing flexibility. Current studies and literature support the idea that both stretching and resistance training independently contribute to improvements in ROM, strength enhancement, and a reduction in the incidence of musculotendinous injuries.

In situations where time constraints or specific circumstances dictate, the benefits of flexibility training can be integrated into resistance training routines. However, it's important to recognize that stretch training still holds its place as a valuable component in overall fitness and training for a broad population. For instance, in scenarios where resistance training might not be suitable as part of a warm-up before competition, stretching can play a pivotal role in preparing the body for specific activities or competitions. Moreover, stretching serves as a relaxation method for many practitioners, where the dynamic and sometimes more intense nature of resistance training might not be as appropriate for achieving a state of relaxation.

While resistance training alone can positively impact flexibility, there are situations where incorporating dedicated stretch training remains beneficial and relevant. The choice between stretching and resistance training can be context-dependent, taking into consideration individual preferences, time constraints, and specific fitness or competition requirements. Both modalities, when appropriately applied, contribute to a well-rounded approach to physical preparation and maintenance. Illustrations of various foam rolling exercises can be found with Figures 12.1–12.13.

Figure 12.1 Dumbbell chest press / flyes with arms extended

Figure 12.2 Dumbbell chest press / flyes (horizontal abduction)

Figure 12.3 Pullover with initial position

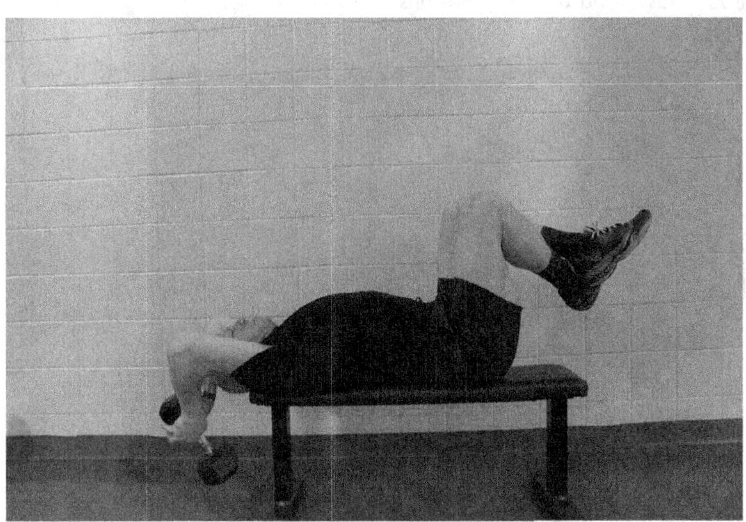

Figure 12.4 Pullover with full shoulder flexion

Figure 12.5 Overhead triceps extensions

Figure 12.6 Lat pulldown with initial position (arms extended)

Figure 12.7 Lat pulldown with final position (shoulder horizontal extension)

Figure 12.8 Seated rowing with elbows extended and scapula protracted

Figure 12.9 Seated rowing with elbows flexed and scapula retracted

Figure 12.10 Incline bicep curls with elbows extended and shoulders in extension

Figure 12.11 Incline bicep curls with elbows flexed

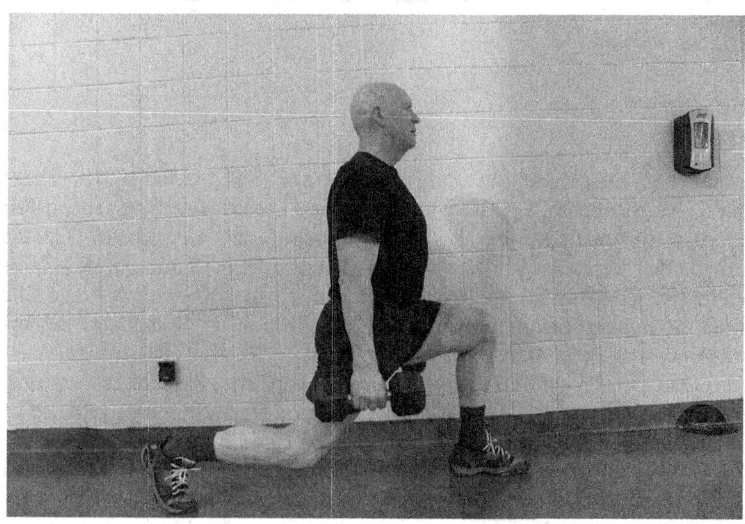

Figure 12.12 Forward lunge

Figure 12.13 Step-up

References

1. Alizadeh, S., Daneshjoo, A., Zahiri, A., Anvar, S.H., Goudini, R., Hicks, J.P., Konrad, A., and Behm, D.G.Resistance Training Induces Improvements in Range of Motion: A Systematic Review and Meta-Analysis. *Sports Med* 53: 707–722, 2023.
2. Behm, D.G., Alizadeh, S., Daneshjoo, A., Anvar, S.H., Graham, A., Zahiri, A., Goudini, R., Edwards, C., Culleton, R., Scharf, C., and Konrad, A.Acute Effects of Various Stretching Techniques on Range of Motion: A Systematic Review with Meta-Analysis. *Sports Med Open* 9: 107, 2023.
3. Behm, D.G., Blazevich, A.J., Kay, A.D., and McHugh, M.Acute effects of muscle stretching on physical performance, range of motion, and injury incidence in healthy active individuals: a systematic review. *Appl Physiol Nutr Metab* 41: 1–11, 2016.
4. Behm, D.G. and Chaouachi, A.A review of the acute effects of static and dynamic stretching on performance. *Eur J Appl Physiol* 111: 2633–2651, 2011.
5. Behm, D.G., Kay, A.D., Trajano, G.S., and Blazevich, A.J.Mechanisms underlying performance impairments following prolonged static stretching without a comprehensive warm-up. *Eur J Appl Physiol* 121: 67–94, 2021.
6. Behm D.G., Kay, A.D., Trajano, G.S., Alizadeh, S., and Blazevich, A.J.Effects of stretching on injury risk reduction and balance. *Journal of Clinical Exercise Physiology* 10: 106–116, 2021.

7. Blazevich, A.J., Cannavan, D., Waugh, C.M., Fath, F., Miller, S.C., and Kay, A.D. Neuromuscular factors influencing the maximum stretch limit of the human plantar flexors. *Journal of applied physiology* 113: 1446–1455, 2012.
8. Bobbert, M.F., Ettema, G.C., and Huijing, P.A.The force-length relationship of a muscle-tendon complex: experimental results and model calculations. *Eur J Appl Physiol Occup Physiol* 61: 323–329, 1990.
9. Fletcher, I.M.The effect of different dynamic stretch velocities on jump performance. *Eur J Appl Physiol* 109: 491–498, 2010.
10. Gardiner, E. *Athletics of the ancient world.* Oxford: Clarendon Press, 1930.
11. Guissard, N. and Duchateau, J.Effect of static stretch training on neural and mechanical properties of the human plantar-flexor muscles. *Muscle and Nerve* 29: 248–255, 2004.
12. Guissard, N. and Duchateau, J.Neural aspects of muscle stretching. *Exerc Sport Sci Rev* 34: 154–158, 2006.
13. Herda, T.J., Herda, N.D., Costa, P.B., Walter-Herda, A.A., Valdez, A.M., and Cramer, J.T.The effects of dynamic stretching on the passive properties of the muscle-tendon unit. *J Sports Sci* 31: 479–487, 2013.
14. Herzog, W. and ter Keurs, H.E.Force-length relation of in-vivo human rectus femoris muscles. *Pflügers Arch* 411: 642–647, 1988.
15. Katz, R. and Pierrot-Deseilligny, E.Recurrent inhibition in humans. *Prog Neurobiol* 57: 325–355, 1999.
16. Kawama, R., Yanase, K., Hojo, T., and Wakahara, T.Acute changes in passive stiffness of the individual hamstring muscles induced by resistance exercise: effects of contraction mode and range of motion. *Eur J Appl Physiol 122*: 2071–2083, 2022.
17. Konrad, A., Stafilidis, S., and Tilp, M.Effects of acute static, ballistic, and PNF stretching exercise on the muscle and tendon tissue properties. *Scand J Med Sci Sports* 27: 1070–1080, 2017.
18. Konrad, A. and Tilp, M.Effects of ballistic stretching training on the properties of human muscle and tendon structures. *J Appl Physiol (1985)* 117: 29–35, 2014.
19. Leighton, J.R.A Study of the Effect of Progressive Weight Training on Flexibility. *J Assoc Phys Ment Rehabil* 18: 101–104 passim, 1964.
20. Lima, C.D., Ruas, C.V., and Behm, D.G., and Brown, L.E.Acute effects of stretching on flexibility and performance: A narrative review. *Journal of Science in Sport and Exercise* 1: 29–37, 2019.
21. Maganaris, C.N.Force-length characteristics of in vivo human skeletal muscle. *Acta Physiol Scand* 172: 279–285, 2001.
22. Magnusson, S.P., Hansen, P., and Kjaer, M.Tendon properties in relation to muscular activity and physical training. *Scand J Med Sci Sports* 13: 211–223, 2003.
23. Magnusson, S.P. and Renstrom, P.The European College of Sports Sciences Position statement: The role of stretching exercises in sports. *European Journal of Sport Science* 6: 87–91, 2006.
24. Magnusson, S.P., Simonsen, E.B., Aagaard, P., Sorensen, H., and Kjaer, M.A mechanism for altered flexibility in human skeletal muscle. *Journal of Physiology* 497 (Pt 1): 291–298, 1996.
25. Mahieu, N.N., McNair, P., De Muynck, M., Stevens, V., Blanckaert, I., Smits, N., and Witvrouw, E.Effect of static and ballistic stretching on the muscle-tendon tissue properties. *Med Sci Sports Exerc* 39: 494–501, 2007.

26. Marusic, J., Vatovec, R., Markovic, G., and Sarabon, N.Effects of eccentric training at long-muscle length on architectural and functional characteristics of the hamstrings. *Scand J Med Sci Sports* 30: 2130–2142, 2020.
27. Mattei, B., Schmied, A., Mazzocchio, R., Decchi, B., Rossi, A., and Vedel, J.P. Pharmacologically induced enhancement of recurrent inhibition in humans: effects on motoneurone discharge patterns. *J Physiol* 548: 615–629, 2003.
28. Reeves, N.D., Maganaris, C.N., Longo, S., and Narici, M.V.Differential adaptations to eccentric versus conventional resistance training in older humans. *Exp Physiol* 94: 825–833, 2009.
29. Scripture, E.W., Smith, T.L., and Brown, E.M.On the education of muscular control and power. *Studies of the Yale Psychology Laboratory* 2: 114–119, 1894.
30. Thomas, E., Ficarra, S., Nakamura, M., Paoli, A., Bellafiore, M., Palma, A., and Bianco, A.Effects of different long-term exercise modalities on tissue stiffness. *Sports Medicine – Open* 8: 71, 2022.

13 Acute and Chronic Effects of Foam Rolling Effects on Range of Motion and Performance

Andreas Konrad

Stretching is not the only technique that can be used to enhance range of motion (ROM). Massage has been used since recorded history as a relaxation technique, cure for specific illnesses and performance enhancement. The first written records of the use of massage originated in Egypt and China. A Chinese text entitled, "The Yellow Emperor's Classic Book of Internal Medicine." was written around 2700 BCE. Egyptian tomb paintings around 2500 BCE illustrate massage as a medical therapy while written records of massage in India originate from 1500 BCE to 500 BCE. Western civilizations did not use massage as medical or health therapy until Per Ling from Sweden incorporated hand stroking into the Swedish Movement System in the early 1800s. Recently, devices such as foam rollers and roller massagers have become quite popular as self-massage therapy (see Figure 13.1) (1, 8, 18, 36, 37, 46, 52).

Fascial Restrictions and Myofascial Adhesions

Originally, foam rolling devices were branded as "self-myofascial release" devices that could aid in reducing (myo)fascial restrictions. Fascial restrictions are reported to occur in response to injury, disease, overuse, inactivity, and inflammation. Purportedly, if fascia loses its elasticity and becomes dehydrated, fascia or connective tissue can bind around the traumatized areas, causing a fibrous adhesion to form (49, 50). These myofascial adhesions may create "hypersensitive tender spots" (58), also known as trigger points. Fibrous adhesions can be painful, prevent normal muscle mechanics (i.e. neuromuscular hypertonicity, and decreased strength, endurance and motor coordination) and decrease soft-tissue extensibility negatively affecting joint ROM and muscle length (49, 50).

Type of foam rollers

Rolling involves small undulations back and forth over the affected muscle with a dense foam roller, typically starting at the proximal portion of the muscle, working down to the distal portion of the muscle or vice versa. Foam rollers are quite diverse in their composition with some made from polyvinyl chloride pipe surrounded by neoprene foam while others may be made from closed cell foam.

DOI: 10.4324/9781032709086-15

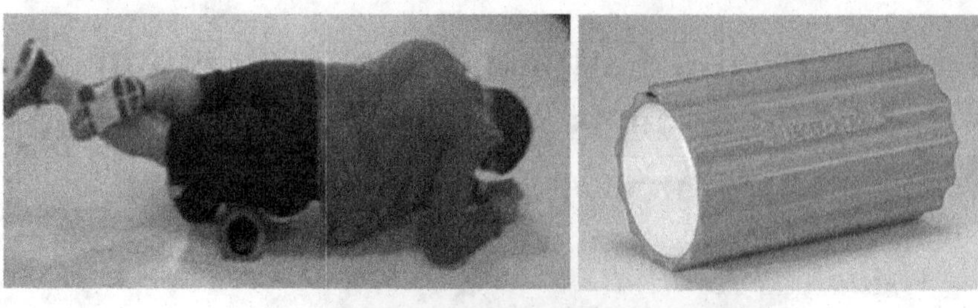

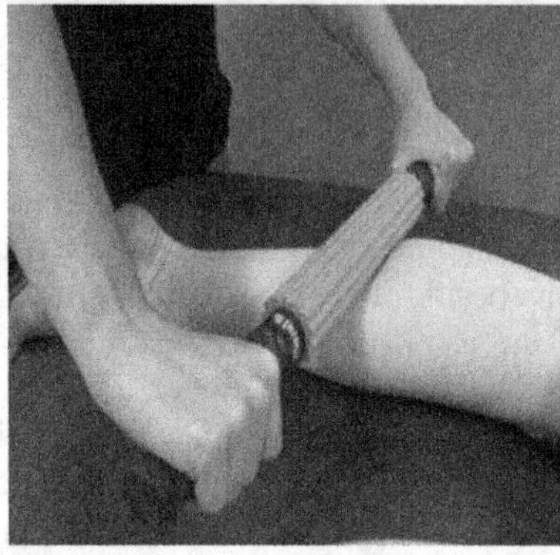

Figure 13.1 Foam rollers and roller massagers

Roller massagers are typically composed of dense foam wrapping around a solid plastic cylinder. There are many variations with some having a ridged design that is supposed to allow for both superficial and deep-tissue massage (8, 52). Besides the cylinder form, there are recent modifications such as foam ball rollers with various sizes, double foam ball rollers specially to treat the back extensors, or even small pointed pyramids to capture deep trigger points. Additionally, most of the devices are able to vibrate and there is a belief that this will increase the impact of the foam rolling exercise (48).

Acute foam rolling effects

A single session of foam rolling can increase the ROM of a joint immediately after the exercise (44, 60), and this increase can last for more than 60 minutes post-exercise (24). The latest techniques such as vibration foam rolling seem to increase the ROM of a joint to a greater extent compared to conventional foam rolling without vibration (45). Regarding the immediate impacts of foam rolling on various performance measures, a meta-analysis (59) suggests a trend towards enhanced sprint performance ($P = 0.06$; +0.7 percent; $ES = 0.28$), with minimal effects on jumping or strength performance. Additionally, reviews such as that by Cheatham et al. (10) corroborate findings indicating that a single bout of foam rolling does not significantly alter performance in the treated body regions. No changes in performance measures following a single bout of foam rolling can actually be a positive response, as one of the major concerns about prolonged static stretching was the evidence for subsequent performance impairments (5, 6). According to meta-analyses there was no significant difference between the acute effects of stretching and foam rolling on ROM (28, 60) as well as on performance parameters (30). However, subgroup analyses revealed positive impacts of foam rolling on performance parameters when contrasted with static stretching, particularly when targeting specific muscles like the quadriceps or specific tasks such as strength exercises, especially when performed for durations exceeding 60 seconds or incorporating vibration. Conversely, when compared to dynamic stretching or utilized without vibration, similar effects were observed with no significant differences.

When combining foam rolling and stretching as a warm-up, a meta-analysis revealed a significant overall effect on ROM of the combined treatment when compared to no intervention (i.e., control condition). However, when a combined treatment (foam rolling + stretching) was compared to foam rolling or stretching alone a similar magnitude of change in ROM was shown (27). This indicates that a time efficient athlete can either stretch or perform foam rolling to acutely increase the ROM, as there seems to be no beneficial effect on ROM if those treatments are combined.

Chronic foam rolling effects

Regarding the long-term (i.e., training, chronic) effects of foam rolling, a recent meta-analysis from our research group (29) revealed that sustained foam rolling practice can lead to increased ROM over time (ES = 0.82), particularly when performed for more than four weeks and focused on the quadriceps and hamstrings. There was no significant improvement with the triceps surae however. One study showed that there is no additional effect when adding a vibration stimulus to the conventional form rolling exercise (23). When it comes to performance parameters, another recent meta-analysis of our laboratory (26) suggests that foam rolling training does not result in significant changes (ES = 0.294; p = 0.281).

Rolling Mechanisms

So how does rolling increase ROM? The more obvious mechanisms exclusively for the acute increase in ROM are warm-up effects or thixotropy. The theory that simple warm-up effects are responsible for the acute changes in ROM is underlined by the literature. Our research group conducted an experiment where we compared a foam rolling exercise for 4x45 seconds on the hamstring muscles with a sham rolling 4x45 seconds by using a roller board to imitate the foam rolling movement. Interestingly, both groups (even the non-rolling sham group) increased the hip extension ROM similarly, which is very likely an indication for a global warm-up effect rather than changes in the local muscle properties (56). Similarly, a meta-analysis reported no beneficial increases in ROM, owing to foam rolling (or stretching) when compared to other warm-up regimes such as cycling or running (57).

Another potential mechanism for an acute increase in ROM following a single bout of foam rolling are thixotropic effects as already discussed earlier. Thixotropic effects occur when viscous (thicker) fluids become less viscous or more fluid-like when agitated, sheared, or stressed. When the stress is removed, or desists then the fluid takes a certain period to return to the original viscous state. The rolling undulations place direct and sweeping pressure on the soft-tissue and generate friction. The increased friction-related temperature of the soft tissues and the stress from the rolling can decrease the viscosity of intracellular and extracellular fluid providing less resistance to movement. Remember the cold car and oil analogy.

The use of the term self-myofascial release suggests that rolling will help to release myofascial adhesions allowing freer less restricted movement. However, it is suggested that the amount of force needed to remove adhesions would exceed the physiological limitations of most people (49, 50). However, the research is not unanimous as it has been shown in one study that a single bout of foam rolling on the anterior thigh acutely released myofascial adhesions (while stretching does not) (34). Up to the present date, there is no evidence, if repeated foam rolling (i.e., training) treatments will release myofascial adhesions. However, as it was reported that a foam rolling training for more than four weeks can increase the ROM (29) one could speculate that at least one mechanism for this increase might be a chronic release in myofascial adhesions. However, are there mechanisms other than thixotropy that might explain changes in ROM following foam rolling?

Golgi tendon organs (GTO: type Ib afferents) responds to musculotendinous tension and strong stretch usually resulting in inhibition. A study from our laboratory (21) demonstrated that using a single bout of a short-duration (ten or 30 seconds) massage at the hamstrings musculotendinous junction increased ROM without an increase in passive muscle tension or electromyography (EMG) activity. Thirty seconds of musculotendinous massage provided greater overall ROM than 10 seconds. While one might assume that the tension applied to the tendon would activate the inhibitory responses of the GTO leading to greater muscle relaxation or decreased tonus, the GTO effects persist for only approximately 60 milliseconds after cessation of stress (54). Thus, the respective 5.8 percent−11.3 percent ROM increases immediately after the massage could not be due to GTO inhibition.

Muscle, fascia and skin are densely innervated by sensory neurons (49, 50). Within the epidermis and dermis (skin layers), there are mechanoreceptors such as the Merkel receptors, Meissner corpuscles, Ruffini cylinders, and Pacinian corpuscles which have different sized receptor fields, adapt slowly or rapidly and respond to different frequencies of skin stimulation. Merkel disks (small receptor field) and Ruffini cylinders (large receptor field) are slowly adapting and respond as long as the stimulus is present whereas the Meissner (small receptor field) and

Pacinian (large receptor field) corpuscles are rapidly adapting and thus respond to stimulation with a burst of firing activity at the beginning and end of stimulation. They respond to a variety of frequencies of stimulation ranging from 0.3–3 Hz (Merkel receptor), 3–40 Hz (Meissner corpuscle), 15–400 Hz (Ruffini cylinder) and 10–500 Hz (Pacinian corpuscle). So, all four mechanoreceptors would respond to slow rolling or massage with Ruffini cylinders and Pacinian corpuscles also responding to high frequency vibrations from machinery. Their major responsibilities are for proprioception and thus would not play a major role in neural inhibition, however Ruffini and Pacinian receptors may be able to contribute to the inhibition of sympathetic activity (contribute to muscle relaxation) (61). Ruffini receptors are more sensitive to tangential forces and lateral stretch (35), which would be prominent with rolling. Ruffini corpuscles stimulation may decrease sympathetic nervous system activity (55). Perhaps the increased relaxation, decreases in heart rate and blood pressure with massage can be partially attributed to contributions from manual stimulation of Ruffini and Pacinian receptors (58).

Another series of receptors that can affect sympathetic and parasympathetic activation are the interstitial type III and IV receptors. These receptors have both low and high threshold sensory capabilities. They are more numerous than type I and II receptors with either having a thin myelinated sheath or are unmyelinated and originate from free nerve endings. They are multi-modal receptors (many functions) responding to pain but also act as mechanoreceptors responding to tension and pressure. Their response to rapid and sustained pressure can contribute to decreases in heart rate, blood pressure, ventilation and lead to vasodilation (41). Thus, they can also contribute to a more relaxed muscle with less resistance to a full range of movement.

In terms of muscle stiffness, there are conflicting reports in the literature, although a current meta-analysis reported no acute changes in muscle stiffness following a single bout of foam rolling (12). Similarly in the long-term, various original studies reported no changes in muscle stiffness following several weeks of foam rolling training (23, 25).

Hence, there is growing evidence that mechanisms of rolling may be due to a number of factors including neural, psychological, and musculotendinous mechanisms. Magnusson has proposed that the acute or immediate increase in ROM following stretching is due to an increased stretch tolerance (38). In other words, the individual while stretching feels discomfort or pain and the act of holding that position allows him/her to diminish or accommodate the painful or uncomfortable sensation. If the same muscle length or joint ROM feels more tolerable then the person should be able to push themselves to an even greater ROM. There could be both psychological and neural components to this effect. Increased pain tolerance after foam rolling was confirmed following a single bout (22) as well as following a foam rolling training (23, 25). Consequently, pain modulation can be considered as a major component for the increase in ROM, owing to foam rolling.

Rolling can diminish different types of pain. Two studies from our lab punished subjects with ten sets of ten squat repetitions. The objective was to induce delayed onset muscle soreness (DOMS). In one study, participants were tested 24, and 48 hours after the squat protocol (46), whereas in the other they were also tested at 72 hours post-exercise (36). One of the groups would roll before testing and the other would not. In both studies, pain perception decreased when rolling was included post-exercise. DOMS-induced decreases in performance measures such as muscle activation, vertical jump, sprint time, and strength-endurance were attenuated (less impairments) with rolling. So, was pain decreased because of self-myofascial release or as a result of central (neural) mechanisms? Subsequent rolling and pain studies from our lab illustrate the importance of the neural component.

Whereas DOMS pain may endure for up to seven days owing to muscle damage and inflammatory responses (4), muscle tender points may persist for weeks and months. Based on the aforementioned studies, rolling can inhibit short duration pain (DOMS) but does it have an effect on chronic pain (i.e., muscle tender points or trigger points)? A study by Aboodarda et al. (1) had a massage therapist identify the most sensitive muscle tender point on participants' plantar flexors (calf). Then the calf was either massaged by the therapist, rolled or not touched (control). One more condition was included and that involved rolling of the contralateral calf. Both massaging and rolling the affected calf decreased the pain sensitivity of the muscle tender point. But even more interesting was that rolling the contralateral calf; *no need to even touch the painful calf*; reduced the pain sensitivity as well. In a similar study (9), we induced pain by electrically stimulating the tibial nerve of the plantar flexors with maximal and submaximal (70 percent of maximal current) intensity high frequency (50 Hz) electrical stimulation (tetanus). The experiment was not much fun for the participants as it involved maximal tolerable pain. Again, we either rolled the affected calf, did nothing (control) or rolled the contralateral calf. While each session of muscle stimulation increased pain sensitivity by about 10 percent, rolling the affected or contralateral calf inhibited the pain increase. Thus, if there is no need to touch the affected muscle and rolling another muscle decreases pain, then it is evident that some global neural response has occurred to decrease pain under chronic (muscle tender points), short term (DOMS) and acute (tetanic stimulation) pain conditions.

In the aforementioned Aboodarda et al. (1) pain depression occurred after light rolling massage. Nociceptors (pain receptors) are present in both muscle and skin (31, 32) and thus even light rolling can increase the sensitivity of superficial nociceptors. Related to this finding, another study from our laboratory (14) reported that the intensity of rolling did not differentially affect the ROM. That is, whether the roller massage was at 50 percent, 70 percent, or 90 percent of the maximum point of discomfort, the increase in ROM was similar.

There are a number of central nervous system possibilities including the gate control theory of pain (see Figure 9.7) (39, 42), diffuse noxious inhibitory control (see Figure 13.2) (40) and parasympathetic nervous system alterations.

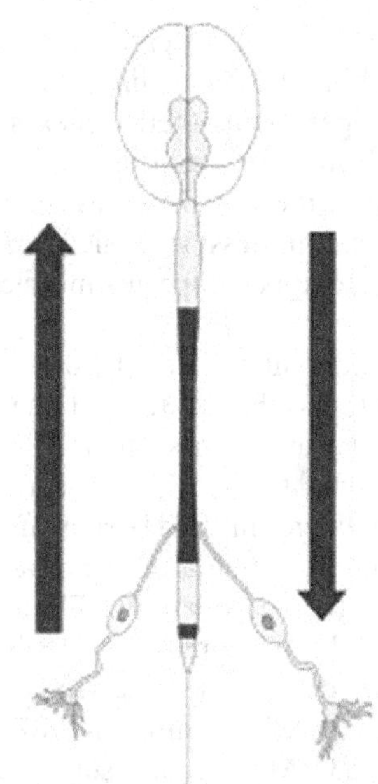

Figure 13.2 Diffuse noxious inhibitory control

The gate control theory of pain is commonly exemplified by somebody hurting a body part and immediately grabbing and rubbing that same body part to minimize the pain. This action would stimulate many receptors and afferents which would then be filtered at the brain. More technically the gate control theory involves activation of thick myelinated ergoreceptor (group III and IV afferents) nerve fibres (via activation of percutaneous [skin] and muscle mechanoreceptors, metaboreceptors and proprioceptors) that modify the signals from ascending nociceptors (pain receptors) via small diameter AΔ fibres to the periaqueductal grey nucleus (42). Analgesia (pain suppression) then results from descending signals to opioid receptors, which would inhibit pain with serotonergic and noradrenergic neurons (47).

Diffuse noxious inhibitory control (DNIC), also known as counter-irritation can be activated by nociceptive stimuli (i.e., mechanical pressure) from a non-local tissue. The typical practical application is that if you banged your head against the cupboard, you can stub (hurt) your toe and the head pain would subjectively decrease. With DNIC, activation of non-local receptors is transmitted to multi-receptive, wide dynamic range convergent neurons in the cortical subnucleus reticularis dorsalis where it inhibits pain transmission monoaminergically (i.e., using monoamine transmitters such as noradrenaline and serotonin) reducing pain perception not only locally but also at distant sites (40, 47, 51).

Massage can also stimulate parasympathetic activation; with changes in serotonin, cortisol, endorphin, and oxytocin contributing to a decreased pain perception (58). Decreased parasympathetic reflexes could decrease pain sensitivity by reducing stress from myofascial tissue by relaxing the strain on the smooth muscles embedded in the soft tissue. Massaging or rolling the muscle can produce acute ischemic compression, which has been shown to result in reduced perceived pain in the upper trapezius muscles (33) and with neck and shoulder muscles (19).

Massage decreases the afferent excitability of the alpha motoneurons. A study from our laboratory found that both tapotement and pettrisage massage (7) as well as roller massage techniques attenuated the H-reflex to M-wave ratio. Normalizing the H-reflex to the M-wave is an important technical procedure to ensure that any changes in the H-reflex are not due to alterations in the muscle's membrane action potential and are actually due to changes in the afferent excitability of the motoneuron. The suppression of the spinal reflex excitability with the H-reflex may be attributed to decreased alpha motoneuron excitability and/or increased presynaptic inhibition. A number of other studies have reported H-reflex inhibition with stretching (2, 11, 15, 16, 17) as well as massage (13, 53). Massage has suppressed the H-reflex by 40–90 percent in these studies.

Thus, while ROM increases with massage and rolling may be attributed to simple warm-up effects or thixotropic factors, a greater tolerance to the discomfort associated with musculotendinous lengthening (stretch tolerance), there could also be reflexive inhibition decreasing stretch-induced contractile activity resulting in a more relaxed muscle. Another potential but controversial mechanism for an increase in ROM might be the release of myofascial adhesions.

Other potential benefits or contradictors

When foam rolling the hamstrings, quadriceps, or calf muscles, it's important to assume a position akin to a plank. A prior study of our laboratory (62) indicated that the muscle engagement, as measured by EMG of the core muscles during a plank or reverse plank position is comparable to that during foam rolling of the quadriceps and hamstrings, respectively. Therefore, it's plausible to suggest that prolonged foam rolling of the front and back thigh muscles, as well as the calves, in a plank-like position could potentially enhance core strength and endurance over time.

Besides all the positive effects proposed in this chapter, potential contradictors have to be discussed. A Delphi study discussed possible reasons to avoid or proceed with caution when using foam rolling, none of which have been documented in clinical trials thus far. The panel of experts identified open wounds and bone fractures as reasons to avoid foam rolling, while conditions such as local tissue inflammation, deep vein thrombosis, myositis ossificans, and osteomyelitis were deemed cautionary. The authors recommended that future research should focus on gathering evidence-based information on when to avoid or use caution with foam rolling for various conditions and in different professional settings (3).

Rolling Recommendations

What intensity and duration of rolling is optimal to achieve the greatest increases in ROM? Studies have demonstrated increased joint ROM with as little as 5–10 seconds of rolling with 10 seconds providing significantly greater increases than 5 seconds (52). Most studies use multiple sets of 30–60 seconds bouts of rolling and although not directly compared 60 seconds rolling seem to provide further ROM improvements (8, 18, 36, 37, 46). However, in one study, 60, 90 and 120 seconds of rolling were applied between four sets of knee extensions. While 120 seconds of rolling decreased the number of knee extension repetitions by 14 percent, the 90 and 60 seconds of rolling also decreased repetition numbers by 8–9 percent (43). This was one of the few studies that has reported impairments with rolling. As mentioned previously, whether rolling was conducted at 5/10, 7/10, or 9/10 on a pain scale, the increases in ROM were similar (14). Thus, there is no need to apply excruciating pain to achieve the desired results. Furthermore, following a typical warm-up (five minutes of aerobic activity, static and dynamic stretching and sport specific activities), if rolling is performed at ten-minute intervals after the warm-up, the augmented ROM is maintained for 30 minutes to a greater degree than without intermitted rolling (20). This finding would be very important for those athletes who do not start a game or match and must sit on a bench for 10–30 minutes before entering the game. Continuing to roll while waiting on the bench would help maintain some of the benefits of the warm-up. Illustrations of various foam rolling exercises can be found with Figures 13.3–13.27.

Figure 13.3 a,b: Gluteals

Figure 13.4 a,b,c: Quadriceps

Figure 13.5 Adductors (Groin)

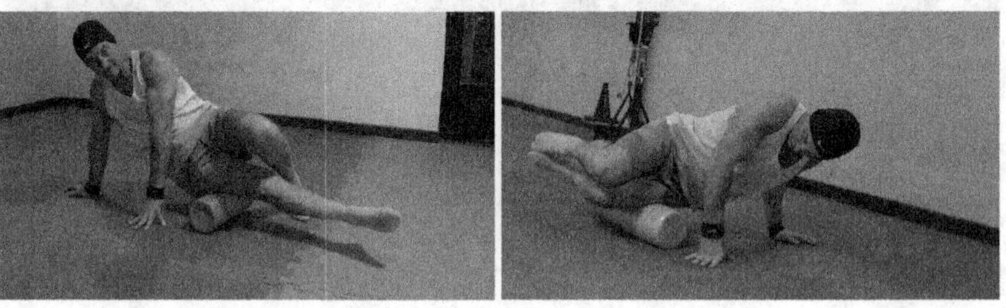

Figure 13.6 a,b: Tensor Fascia Latae

Figure 13.7 a,b: Hamstrings

Figure 13.8 Plantar Flexors

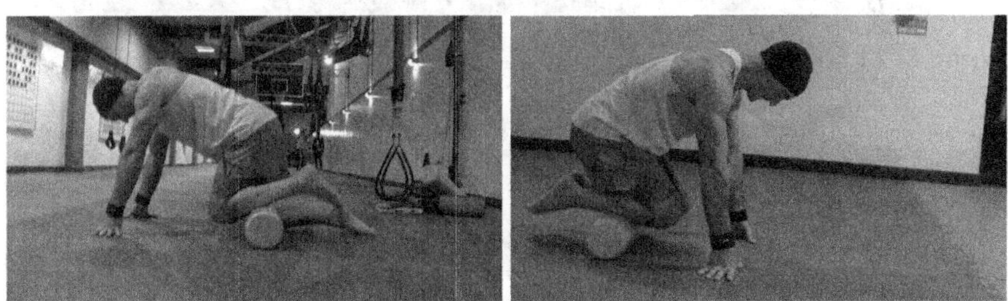

Figure 13.9 a,b: Tibialis Anterior

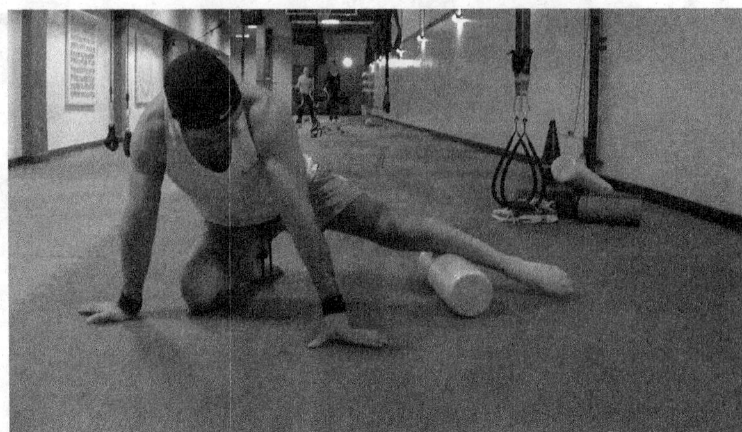

Figure 13.10 Inner calf

Figure 13.11 Outer calf

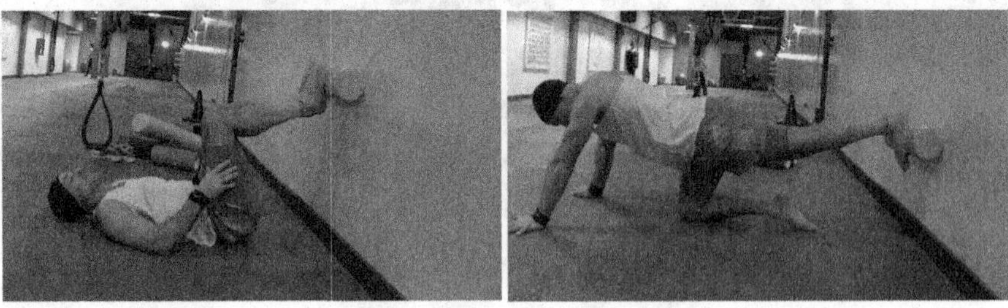

Figure 13.12 a,b: Sole of the foot

Figure 13.13 Pectorals

Figure 13.14 Pectorals and Anterior Deltoid

Figure 13.15 Lateral Shoulder (Deltoid) and Arm (Biceps and Triceps Brachii)

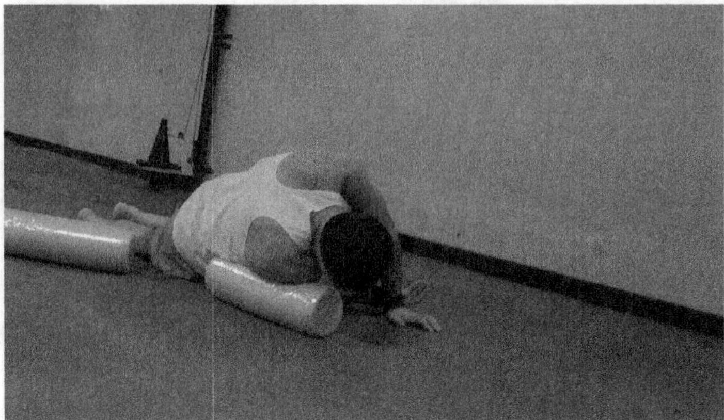

Figure 13.16 Posterior Shoulder (Deltoids) and Triceps Brachii

Figure 13.17 Posterior Shoulders (Deltoids and Trapezius)

Figure 13.18 Biceps Brachii

Figure 13.19 Triceps Brachii

Figure 13.20 Back (Upright position)

Figure 13.21 Lower Back

Figure 13.22 a,b: Lateral Trunk

Figure 13.23 Neck

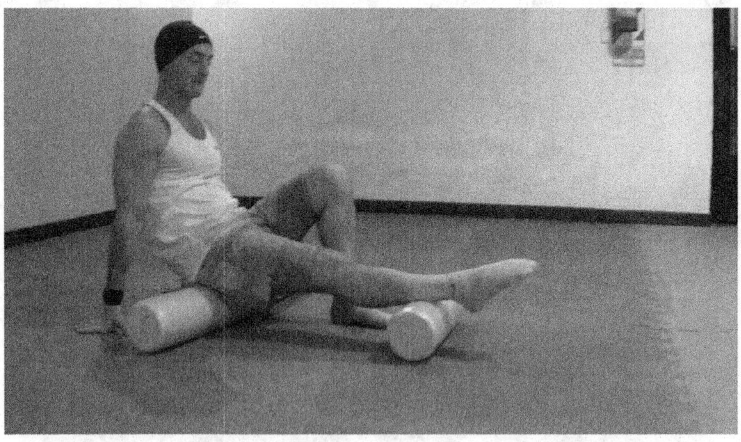

Figure 13.24 Gluteals and Plantar Flexors (Double Rolling)

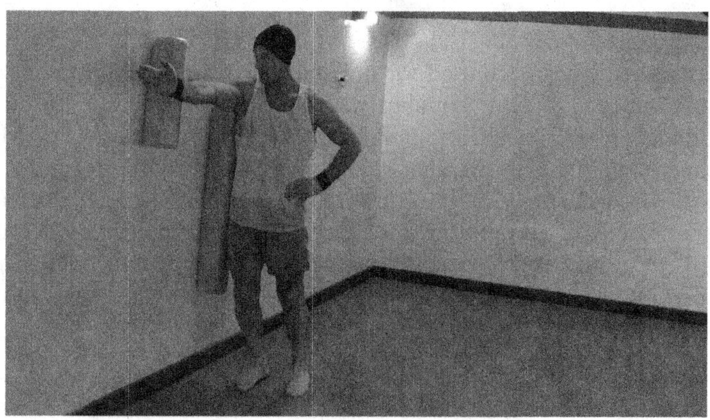

Figure 13.25 Back and Arm (Double Rolling)

Figure 13.26 Roller Massage of the hamstrings.

Figure 13.27 Exercise Balls

Summary

Whereas massage has been used for millennia, foam rollers and roller massagers have enjoyed recent popularity. Rolling can increase flexibility for approximately 60 minutes without subsequent performance deficits. Mechanisms underlying these ROM changes may be ascribed to warm-up effects, decreased visco-elasticity (thixotropic effects), rolling-induced decreases in sympathetic activity and motoneuron excitability and increased stretch tolerance. It is recommended to perform multiple sets of 30–60 seconds of rolling, which can be performed below the maximum pain tolerance (50–90 percent of maximum pain tolerance).

References

1. Aboodarda, S.J., Spence, A.J., and Button, D.C.Pain pressure threshold of a muscle tender spot increases following local and non-local rolling massage. *BMC Musculoskelet Disord* 16: 265, 2015.
2. Avela, J., Kyröläinen, H., and Komi, P.V.Altered reflex sensitivity after repeated and prolonged passive muscle stretching. *Journal of Applied Physiology* 86: 1283–1291, 1999.
3. Bartsch, K.M., Baumgart, C., Freiwald, J., Wilke, J., Slomka, G., Turnhofer, S., Egner, C., Hoppe, M.W., Klingler, W., and Schleip, R.Expert Consensus on the Contraindications and Cautions of Foam Rolling-An International Delphi Study. *J Clin Med* 10, 2021.
4. Behm, D.G., Baker, K.M., Kelland, R., and Lomond, J.The effect of muscle damage on strength and fatigue deficits. *J Strength Cond Res* 15: 255–263, 2001.
5. Behm, D.G., Blazevich, A.J., Kay, A.D., and McHugh, M.Acute effects of muscle stretching on physical performance, ROM, and injury incidence in healthy active individuals: a systematic review. *Appl Physiol Nutr Metab* 41: 1–11, 2016.
6. Behm, D.G. and Chaouachi, A.A review of the acute effects of static and dynamic stretching on performance. *Eur J Appl Physiol* 111: 2633–2651, 2011.
7. Behm, D.G., Peach, A., Maddigan, M., Aboodarda, S.J., DiSanto, M.C., Button, D.C., and Maffiuletti, N.A.Massage and stretching reduce spinal reflex excitability without affecting twitch contractile properties. *J Electromyogr Kinesiol* 23: 1215–1221, 2013.
8. Bradbury-Squires, D.J., Noftall, J.C., Sullivan, K.M., Behm, D.G., Power, K.E., and Button, D.C.Roller-massager application to the quadriceps and knee-joint ROM and neuromuscular efficiency during a lunge. *J Athl Train* 50: 133–140, 2015.
9. Cavanaugh, M.T., Döweling, A., Young, J.D., Quigley, P.J., Hodgson, D.D., Whitten, J.H., Reid, J.C., Aboodarda, S.J., and Behm, D.G.An acute session of roller massage prolongs voluntary torque development and diminishes evoked pain. *Eur J Appl Physiol*, 2016.
10. Cheatham, S.W., Kolber, M.J., Cain, M., and Lee, M.The Effects of Self-Myofascial Release Using a Foam Roll or Roller Massager on Joint ROM, Muscle Recovery, and Performance: A Systematic Review. *Int J Sports Phys Ther* 10: 827–838, 2015.
11. Etnyre, B.R. and Abraham, L.D.H-reflex changes during static stretching and two variations of proprioceptive neuromuscular facilitation techniques. *Electroencephalography Clinical Neurophysiology* 63: 174–179, 2005.

12. Glanzel, M.H., Rodrigues, D.R., Petter, G.N., Pozzobon, D., Vaz, M.A., and Geremia, J.M.Foam Rolling Acute Effects on Myofascial Tissue Stiffness and Muscle Strength: A Systematic Review and Meta-Analysis. *J Strength Cond Res* 37: 951–968, 2023.
13. Goldberg, J., Sullivan, S.J., and Seaborne, D.E.The effect of two intensities of massage on H-reflex amplitude. *Phys Ther* 72: 449–457, 1992.
14. Grabow, L., Young, J.D., Alcock, L.R., Quigley, P.J., Byrne, J.M., Granacher, U., Skarabot, J., and Behm, D.G.Higher Quadriceps Roller Massage Forces Do Not Amplify Range-of-Motion Increases nor Impair Strength and Jump Performance. *J Strength Cond Res* 32: 3059–3069, 2018.
15. Guissard, N. and Duchateau, J.Neural aspects of muscle stretching. *Exerc Sport SciRev* 34: 154–158, 2006.
16. Guissard, N., Duchateau, J., and Hainaut, K.Muscle stretching and motoneuron excitability. *European Journal of Applied Physiology* 58: 47–52, 1988.
17. Guissard, N., Duchateau, J., and Hainaut, K.Mechanisms of decreased motoneurone excitation during passive muscle stretching. *Experimental Brain Research* 137: 163–169, 2001.
18. Halperin, I., Aboodarda, S.J., Button, D.C., Andersen, L.L., and Behm, D.G. Roller massager improves ROM of plantar flexor muscles without subsequent decreases in force parameters. *Int J Sports Phys Ther* 9: 92–102, 2014.
19. Hanten, W.P., Olson, S.L., Butts, N.L., and Nowicki, A.L.Effectiveness of a home program of ischemic pressure followed by sustained stretch for treatment of myofascial trigger points. *Phys Ther* 80: 997–1003, 2000.
20. Hodgson, D.D., Quigley, P.J., Whitten, J.H.D., Reid, J.C., and Behm, D.G.Impact of 10-Minute Interval Roller Massage on Performance and Active ROM. *Journal of Strength and Conditioning Research*, 2017.
21. Huang, S.Y., Di Santo, M., Wadden, K.P., Cappa, D.F., Alkanani, T., and Behm, D.G.Short-duration massage at the hamstrings musculotendinous junction induces greater ROM. *J Strength Cond Res* 24: 1917–1924, 2010.
22. Kasahara, K., Konrad, A., Yoshida, R., Murakami, Y., Koizumi, R., Sato, S., Ye, X., Thomas, E., and Nakamura, M.Comparison of the Prolonged Effects of Foam Rolling and Vibration Foam Rolling Interventions on Passive Properties of Knee Extensors. *J Sports Sci Med* 21: 580–585, 2022.
23. Kasahara, K., Konrad, A., Yoshida, R., Murakami, Y., Sato, S., Aizawa, K., Koizumi, R., Thomas, E., and Nakamura, M.Comparison between 6-week foam rolling intervention program with and without vibration on rolling and non-rolling sides. *Eur J Appl Physiol* 122: 2061–2070, 2022.
24. Kasahara, K., Konrad, A., Yoashida, R., Murakami, Y., Koizumi, R., Thomas, E., and Nakamura, M.Comparison of the effects of different foam rolling durations on knee extensors function. *Biology of Sport* 41: 139–145, 2023.
25. Kiyono, R., Onuma, R., Yasaka, K., Sato, S., Yahata, K., and Nakamura, M. Effects of 5-Week Foam Rolling Intervention on ROM and Muscle Stiffness. *J Strength Cond Res*, 2020.
26. Konrad, A., Nakamura, M., and Behm, D.G.The Effects of Foam Rolling Training on Performance Parameters: A Systematic Review and Meta-Analysis including Controlled and Randomized Controlled Trials. *Int J Environ Res Public Health* 19, 2022.
27. Konrad, A., Nakamura, M., Bernsteiner, D., and Tilp, M.The Accumulated Effects of Foam Rolling Combined with Stretching on ROM and Physical Performance: A Systematic Review and Meta-Analysis. *J Sports Sci Med* 20: 535–545, 2021.

28. Konrad, A., Nakamura, M., Paternoster, F.K., Tilp, M., and Behm, D.G.A comparison of a single bout of stretching or foam rolling on ROM in healthy adults. *Eur J Appl Physiol* 122: 1545–1557, 2022.
29. Konrad, A., Nakamura, M., Tilp, M., Donti, O., and Behm, D.G.Foam rolling training effects on ROM: A Systematic Review and Meta-Analysis. *Sports Med* 52: 2523–2535, 2022.
30. Konrad, A., Tilp, M., and Nakamura, M.A Comparison of the Effects of Foam Rolling and Stretching on Physical Performance. A Systematic Review and Meta-Analysis. *Front Physiol* 12: 720531, 2021.
31. Kosek, E., Ekholm, J., and Hansson, P.Increased pressure pain sensibility in fibromyalgia patients is located deep to the skin but not restricted to muscle tissue. *Pain* 63: 335–339, 1995.
32. Kosek, E., Ekholm, J., and Hansson, P.Pressure pain thresholds in different tissues in one body region. The influence of skin sensitivity in pressure algometry. *Scand J Rehabil Med* 31: 89–93, 1999.
33. Kostopoulos DN, J., Arthur, J., Ingber, R.S., and Larkin, R.W.Reduction of spontaneous electrical activity and pain perception of trigger points in the upper trapezius muscle through trigger point compression and passive stretching. *Journal of Musculoskeletal Pain* 16: 266–278, 2008.
34. Krause, F., Wilke, J., Niederer, D., Vogt, L., and Banzer, W.Acute effects of foam rolling on passive stiffness, stretch sensation and fascial sliding: A randomized controlled trial. *Hum Mov Sci* 67: 102514, 2019.
35. Kruger, L. *Cutaneous Sensory System*. Boston, MA: Birkhauser, 1987.
36. Macdonald, G.Z., Button, D.C., Drinkwater, E.J., and Behm, D.G.Foam rolling as a recovery tool after an intense bout of physical activity. *Med Sci Sports Exerc* 46: 131–142, 2014.
37. MacDonald, G.Z., Penney, M.D., Mullaley, M.E., Cuconato, A.L., Drake, C.D., Behm, D.G., and Button, D.C.An acute bout of self-myofascial release increases ROM without a subsequent decrease in muscle activation or force. *J Strength Cond Res* 27: 812–821, 2013.
38. Magnusson, S.P., Simonsen, E.B., Aagaard, P., Sorensen, H., and Kjaer, M.A mechanism for altered flexibility in human skeletal muscle. *Journal of Physiology* 497 (Pt 1): 291–298, 1996.
39. Melzack, R. and Wall, P.D.Pain mechanisms: a new theory. *Science* 150: 971–979, 1965.
40. Mense, S.Neurobiological concepts of fibromyalgia–the possible role of descending spinal tracts. *Scand J Rheumatol Suppl* 113: 24–29, 2000.
41. Mitchell, J.H. and Schmidt, R.F.Cardiovascular reflex control by afferent fibers from skeletal muscle receptors. Bethesda, MA: American Physiological Society, 1977.
42. Moayedi, M. and Davis, K.D.Theories of pain: from specificity to gate control. *J Neurophysiol* 109: 5–12, 2013.
43. Monteiro, E.R., Vigotsky, A., Skarabot, J., Brown, A.F., del Melo Fiuza, A.G.F., Gomes, T.M., Halperin, I., and da Silva Novaes, J.Acute effects of different foam rolling volumes in the interset rest period on maximum repetition performance. *Hong Kong Physiotherapy Journal* 36: 57–62, 2017.
44. Nakamura, M., Onuma, R., Kiyono, R., Yasaka, K., Sato, S., Yahata, K., Fukaya, T., and Konrad, A.The Acute and Prolonged Effects of Different Durations of Foam Rolling on ROM, Muscle Stiffness, and Muscle Strength. *J Sports Sci Med* 20: 62–68, 2021.

45. Park, S.J., Lee, S.I., Jeong, H.J., and Kim, B.G.Effect of vibration foam rolling on the ROM in healthy adults: a systematic review and meta-analysis. *J Exerc Rehabil* 17: 226–233, 2021.
46. Pearcey, G.E., Bradbury-Squires, D.J., Kawamoto, J.E., Drinkwater, E.J., Behm, D.G., and Button, D.C.Foam rolling for delayed-onset muscle soreness and recovery of dynamic performance measures. *J Athl Train* 50: 5–13, 2015.
47. Pud, D., Granovsky, Y., and Yarnitsky, D.The methodology of experimentally induced diffuse noxious inhibitory control (DNIC)-like effect in humans. *Pain* 144: 16–19, 2009.
48. Reiner, M.M., Glashuttner, C., Bernsteiner, D., Tilp, M., Guilhem, G., Morales-Artacho, A., and Konrad, A.A comparison of foam rolling and vibration foam rolling on the quadriceps muscle function and mechanical properties. *Eur J Appl Physiol* 121: 1461–1471, 2021.
50. Schleip, R.Fascial plasticity – a new neurobiological explanation: Part 1. *Journal of Bodywork and Movement Therapies* 7: 11–19, 2003.
49. Schleip, R.Fascial plasticity – a new neurobiological explanation: Part 2. *Journal of Bodywork and Movement Therapies* 7: 104–116, 2003.
51. Sigurdsson, A. and Maixner, W.Effects of experimental and clinical noxious counterirritants on pain perception. *Pain* 57: 265–275, 1994.
52. Sullivan, K.M., Silvey, D.B., Button, D.C., and Behm, D.G.Roller-massager application to the hamstrings increases sit-and-reach ROM within five to ten seconds without performance impairments. *Int J Sports Phys Ther* 8: 228–236, 2013.
53. Sullivan, S.J., Williams, L.R., Seaborne, D.E., and Morelli, M.Effects of massage on alpha motoneuron excitability. *Phys Ther* 71: 555–560, 1991.
54. Trajano, G.S., Nosaka, K., and Blazevich, A.J.Neurophysiological Mechanisms Underpinning Stretch-Induced Force Loss. *Sports Med*, 2017.
55. van den Berg, F. and Cabri, J. *Angewandte Physiologie – Das Bindegewebe des Bewegungsapparates verstehen und beeinflussen.* Stuttgart: Georg Thieme Verlag, 1999.
56. Warneke, K., Aragao-Santos, J.C., Alizadeh, S., Bahrami, M., Anvar, S.H., Konrad, A., and Behm, D.G.Are Acute Effects of Foam-Rolling Attributed to Dynamic Warm Up Effects? A Comparative Study. *J Sports Sci Med* 22: 180–188, 2023.
57. Warneke, K., Ploschberger, G., Lohmann, L.H., Lichtenstein, E., Jochum, D., Siegel, S.D., Zech, A., and Behm, D.G.Foam rolling and stretching do not provide superior acute flexibility and stiffness improvements compared to any other warm-up intervention: A systematic review with meta-analysis. *J Sport Health Sci*, 2024.
58. Weerapong, P., Hume, P.A., and Kolt, G.S.The Mechanisms of Massage and Effects on Performance, Muscle Recovery and Injury Prevention. *Sports Medicine* 35: 235–256, 2005.
59. Wiewelhove, T., Döweling, A., Schneider, C., Hottenrott, L., Meyer, T., Kellmann, M., Pfeiffer, M., and Ferrauti, A.A Meta-Analysis of the Effects of Foam Rolling on Performance and Recovery. *Frontiers of Physiology* 10, 2019.
60. Wilke, J., Muller, A.L., Giesche, F., Power, G., Ahmedi, H., and Behm, D.G.Acute Effects of Foam Rolling on ROM in Healthy Adults: A Systematic Review with Multilevel Meta-analysis. *Sports Med* 50: 387–402, 2020.
61. Wu, G., Ekedahl, R., Stark, B., Carlstedt, T., Nilsson, B., and Hallin, R.G.Clustering of Pacinian corpuscle afferent fibres in the human median nerve. *Experimental Brain Research* 126: 399–409, 1999.
62. Zahiri, A., Alizadeh, S., Daneshjoo, A., Pike, N., Konrad, A., and Behm, D.G. Core Muscle Activation With Foam Rolling and Static Planks. *Front Physiol* 13: 852094, 2022.

14 Local Vibration Effects on Range of Motion and Performance

Effects on Range of Motion (ROM)

Local muscle vibration (see Figure 14.1) alone and combined with static stretching have been used to enhance range of motion (ROM) improvements. Vibration (35 Hz with 2 mm amplitude displacement) and static stretching have increased hamstrings flexibility more than static stretching (7.8 percent) alone (10), while the combination has also improved the ability to perform the forward splits with male gymnasts (19). With synchronized swimmers, vibration (30 Hz at 2 mm displacement) improved passive but not active ROM (forward splits) (18). Local vibration (30 Hz at 4 mm displacement) alone demonstrated similar stand and reach ROM (lower back and

Figure 14.1 Local vibration device

DOI: 10.4324/9781032709086-16

hamstrings) increases as static stretching with better results than dynamic stretching and whole body vibration (14). There were no statistically significant differences compared to static stretching but local vibration-induced ROM was consistently higher with moderate effect size magnitudes. The vibration effects were especially effective with the most flexible participants (14). An early study from 1976 used 15 minutes of vibration to the low back and hamstrings (44 Hz with 0.1-mm displacements) and reported similar increases in ROM as static stretching (1). Their vibration displacement was very small which could have been the difference between their study and other studies that did show a greater benefit with vibration. Two studies vibrating the plantar flexors found a large magnitude increase in dorsiflexion ROM with no changes in plantar flexor MVC torque (13) while the other study showed similar increases in ankle ROM when comparing plantar flexors static stretching to vibration and static stretching combined (15). Local vibration of the quadriceps (30 Hz vibration with 10 mm amplitude for ten seconds with five seconds recovery) increased knee ROM immediately after and ten minutes after the vibration intervention, whether male American football players had prior knee injuries or not (12).

Although five sessions of local vibration with a Theragun significantly improved active ankle dorsiflexion ROM, Instrument Assisted Soft Tissue Mobilization (IASTM) provided greater ROM increases (21). The authors suggested that IASTM penetrates the muscle deeper than Theragun local vibration providing better fibroblast and fibroclast cell formation and collagen fibres remodeling. A recent systematic review of 11 studies (7) reported that an acute session of percussive massage devices could be effective for increasing iliopsoas, hamstrings, triceps surae and posterior chain muscle flexibility.

Effects on Performance

Although, the improved flexibility with vibration is relatively consistent, research reports on its effect on subsequent performance is mixed. Vibration has counteracted static stretch-induced impairments in jump height (6), and knee flexor and extensor forces (10), in two studies but demonstrated that vibration alone or in combination with static stretching still induced impairments in strength and activation (8, 15) in two other studies. The previously cited Ferreira et al. systematic review (7) reported a lack of improvements in strength, balance, acceleration, agility, and explosive activities.

Mechanisms

The mechanisms underlying the increased ROM with vibration have been attributed to a myriad of factors (18). The factors include increased stretch (pain) threshold, increased blood flow associated with an increase in temperature or vibration stress (11, 16, 17), which would decrease muscle viscosity reducing the resistance to muscle extensibility. Increases in passive straight leg raise ROM following 2 x 60 seconds of handheld percussive therapy (29 Hz) with 30-seconds of rest between the repetitions induced an increase in hamstrings ROM in association with a resultant decrease in tissue stiffness (20). Fifteen minutes of back percussive massage (30 Hz) did not increase lumbar (lower back) flexibility or affect thoracolumbar fascia thickness. But there were decreases in echo intensity and perceived stiffness and an elevated skin temperature (23). A decrease in echo intensity is purported to reflect muscle quality, owing to an attenuation of intramuscular fibrous and adipose tissue. The authors suggested that the decreased echo intensity may have been related to a reduction in the viscosity of hyaluronic acid between the loose connective tissue. Vibration-induced muscle relaxation (9, 22), and decreases in the phasic and static stretch reflexes (2) have also been reported. With even greater nuance, high frequency (90 Hz) vibration affected short-latency responses more than medium-latency responses, but when the vibration frequency was reduced to 30 Hz there was negligible effect on the short-latency reflex, but the medium-latency reflex was significantly reduced (4). Bove and colleagues (5) suggested that these effects were due to pre-synaptic inhibition of the group Ia afferent fibres or a gate control type limitation where both vibration and stretching influence the same Ia pathways. Furthermore, if the stress is high enough, the combination of a strong stretch stimulus and substantial vibration could activate Golgi tendon organ inhibition through the Ib pathways (autogenic inhibition). However, Golgi tendon organ inhibition only persists for less than a second after the stimulus is removed. Additionally, other factors might include intrafusal fibre fatigue or persistence of motoneuron after-discharge, owing to reverberation of the interneuron pool (3). In summary, percussive massage may elicit a number of ROM enhancing mechanisms such as the ubiquitous increase in stretch/pain tolerance, increased blood flow and tissue temperature promoting a decrease in tissue viscosity (thixotropic effects). There also seems to be an attenuation of various reflexes but the post-vibration duration of these effects would be very brief.

Summary

So, adding vibration to the stretching routine is probably not necessary for the average individual interested in improving this musculoskeletal health factor (flexibility), as stretching on its own should accomplish that objective. Alternatively, the research indicates that local vibration in many cases is as effective as stretching for improving ROM. Of course, you do not have to spend money on a percussive massage device if you just stretch. However, for those individuals who need to go further and achieve more extreme ROM, then vibration in concert with stretching might be recommended.

References

1. Atha, J. and Wheatley, D.W.Joint mobility changes due to low frequency vibration and stretching exercise. *Br J Sports Med* 10: 26–34, 1976.

2. Bishop, B.Vibratory stimulation, Part I. Neurophysiology or motor responses evoked by vibratory stimulation. *Physical Therapy* 54: 1273–1282, 1974.
3. Bongiovanni, L.G. and Hagbarth, K.E.Tonic vibration reflexes elicited during fatigue from maximal voluntary contractions in man. *J Physiol* 423: 1–14, 1990.
4. Bove, M., Nardone, A., and Schieppati, M.Effects of leg muscle tendon vibration on group Ia and group II reflex responses to stance perturbation in humans. *J Physiol* 550: 617–630, 2003.
5. Claus, D., Mills, K.R., and Murray, N.M.The influence of vibration on the excitability of alpha motoneurones. *Electroencephalogr Clin Neurophysiol* 69: 431–436, 1988.
6. Fernandes, I.A., Kawchuk, G., Bhambhani, Y., and Gomes, P.S.Does vibration counteract the static stretch-induced deficit on muscle force development? *J Sci Med Sport* 16: 472–476, 2013.
7. Ferreira, R.M., Silva, R., Vigario, P., Martins, P.N., Casanova, F., Fernandes, R.J., and Sampaio, A.R.The Effects of Massage Guns on Performance and Recovery: A Systematic Review. *J Funct Morphol Kinesiol* 8, 2023.
8. Herda, T.J., Ryan, E.D., Smith, A.E., Walter, A.A., Bemben, M.G., Stout, J.R., and Cramer, J.T.Acute effects of passive stretching vs vibration on the neuromuscular function of the plantar flexors. *Scand J Med Sci Sports* 19: 703–713, 2009.
9. Issurin, V.B., Liebermann, D.G., and Tenenbaum, G.Effect of vibratory stimulation training on maximal force and flexibility. *J Sports Sci* 12: 561–566, 1994.
10. Jemni, M., Mkaouer, B., Marina, M., Asllani, A., and Sands, W.A.Acute static vibration-induced stretching enhanced muscle viscoelasticity but did not affect maximal voluntary contractions in footballers. *J Strength Cond Res* 28: 3105–3114, 2014.
11. Kerschan-Schindl, K., Grampp, S., Henk, C., Resch, H., Preisinger, E., Fialka-Moser, V., and Imhof, H.Whole-body vibration exercise leads to alterations in muscle blood volume. *Clin Physiol* 21: 377–382, 2001.
12. Klimowska, N, Jaskulski, K., and Zdrodowska, A.The influence of percussion massage on knee's range of motion in two positions. *Biomedical Human Kinetics* 15: 181–184, 2023.
13. Konrad, A., Glashuttner, C., Reiner, M.M., Bernsteiner, D., and Tilp, M.The Acute Effects of a Percussive Massage Treatment with a Hypervolt Device on Plantar Flexor Muscles' Range of Motion and Performance. *J Sports Sci Med* 19: 690–694, 2020.
14. Kurt, C.Alternative to traditional stretching methods for flexibility enhancement in well-trained combat athletes: local vibration versus whole-body vibration. *Biol Sport* 32: 225–233, 2015.
15. Miller, J.D., Herda, T.J., Trevino, M.A., and Mosier, E.M.The effects of passive stretching plus vibration on strength and activation of the plantar flexors. *Appl Physiol Nutr Metab* 41: 917–923, 2016.
16. Needs, D., Blotter, J., Cowan, M., Fellingham, G., Johnson, A.W., and Feland, J.B. Effect of Localized Vibration Massage on Popliteal Blood Flow. *J Clin Med* 12, 2023.
17. Rittweger, J., Beller, G., and Felsenberg, D.Acute physiological effects of exhaustive whole-body vibration exercise in man. *Clin Physiol* 20: 134–142, 2000.
18. Sands, W.A., McNeal, J.R., Stone, M.H., Kimmel, W.L., Haff, G.G., and Jemni, M.The effect of vibration on active and passive range of motion in elite female synchronized swimmers. *European Journal of Sport Science* 8: 217–223, 2008.
19. Sands, W.A., McNeal, J.R., Stone, M.H., Russell, E.M., and Jemni, M.Flexibility enhancement with vibration: Acute and long-term. *Med Sci Sports Exerc* 38: 720–725, 2006.

20. Skinner, B., Dunn, L., and Moss, R.The Acute Effects of Theragun Percussive Therapy on Viscoelastic Tissue Dynamics and Hamstring Group Range of Motion. *J Sports Sci Med* 22: 496–501, 2023.
21. Tyagi, A.Effect of IASTM vs. THERAGUN on Triceps Surae Active Range of Motion and Functional Movements in University Level Sprinters. *Indian Journal of Physiotherapy and Occupational Therapy* 16: 1–6, 2022.
22. van den Tillaar, R.Will whole-body vibration training help increase the range of motion of the hamstrings? *J Strength Cond Res* 20: 192–196, 2006.
23. Yang, C., Huang, X., Li, Y., Sucharit, W., Sirasaporn, P., and Eungpinichpong, W. Acute Effects of Percussive Massage Therapy on Thoracolumbar Fascia Thickness and Ultrasound Echo Intensity in Healthy Male Individuals: A Randomized Controlled Trial. *International Journal of Environmental Research and Public Health* 20: 1073–1080, 2023.

15 Instrument-Assisted Soft Tissue Mobilization

Another treatment that is gaining popularity is Instrument Assisted Soft Tissue Mobilization (IASTM). Although recently popular, its origins were many millennia in the past and related to Gua Sha, which was a scraping technique leading to light bruising (40). IASTM is similar to manual massage or foam rolling as it involves placing stress and strain on the tissues. This is accomplished with metal tools that conform to specific body areas and permit individuals to apply pressure. It is suggested that IASTM allows more direct treatment of tissues than activities like stretching. Similar to the rationale presented for early foam rolling articles, IASTM is purported to be a soft tissue treatment technique that can positively reduce scar tissue and myofascial adhesions. However, again similar to the lack of foam rolling proof (9), the evidence for attenuating or removing myofascial restrictions or trigger points with IASTM is lacking. Whereas some of the purported mechanisms are not clear, the IASTM effects on ROM have been established.

Acute Effects on Range of Motion

There are a variety of studies demonstrating improved range of motion (ROM) with a single session (acute) of IASTM. IASTM-induced increases in ROM have been recorded at the neck (cervical) (1, 35), with shoulder internal rotation (2, 31) and horizontal adduction (31), hamstrings (hip flexion) (13, 30, 39, 41), knee (45) and hip (37), as well as ankle dorsiflexion (19, 33, 44). These ROM improvements were exhibited after IASTM treatment durations of as little as two (37) and five minutes (13, 30). Shoulder internal rotation ROM improvements with IASTM endured 45 minutes after treatment, which was significantly longer than with than kinesiology taping (2). One study reported ROM gains were still significant 24 hours after IASTM treatment (37).

A number of these studies showed similar ROM benefits when employing IASTM compared to foam rolling of the cervical area (1), hip, and knee (37, 45), shoulder internal rotation with flossing (2), therapeutic cupping (13), vibration massage, light hand massage (30), muscle energy technique with the hamstrings (39), stretching for hip flexion active ROM (41), and dorsiflexion

DOI: 10.4324/9781032709086-17

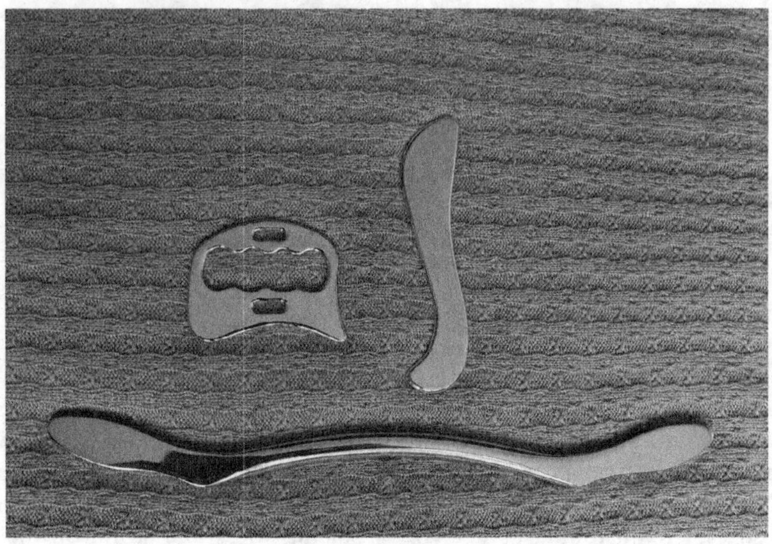

Figure 15.1 Instrument Assisted Soft Tissue Mobilization (IASTM) devices

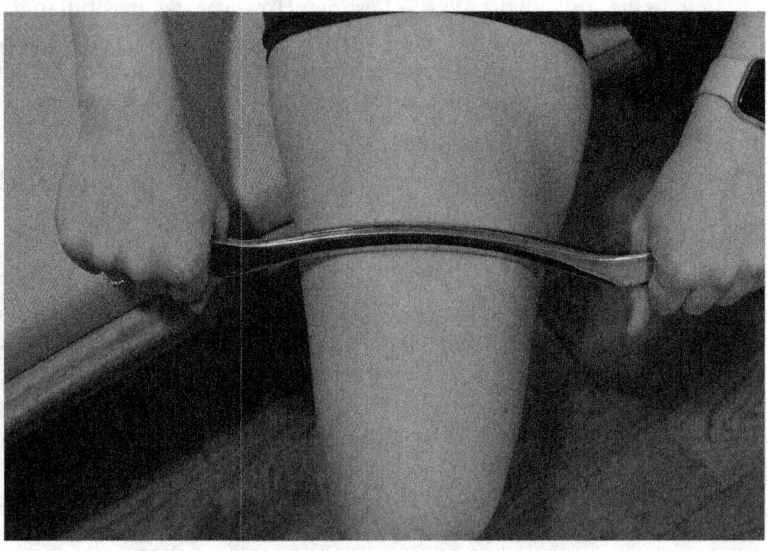

Figure 15.2 Instrument Assisted Soft Tissue Mobilization (IASTM) use on quadriceps

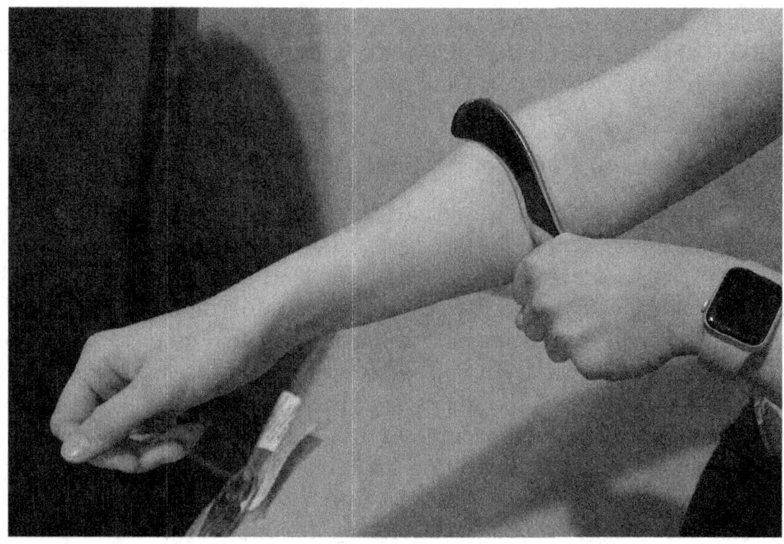

Figure 15.3 Instrument Assisted Soft Tissue Mobilization (IASTM) use on forearms

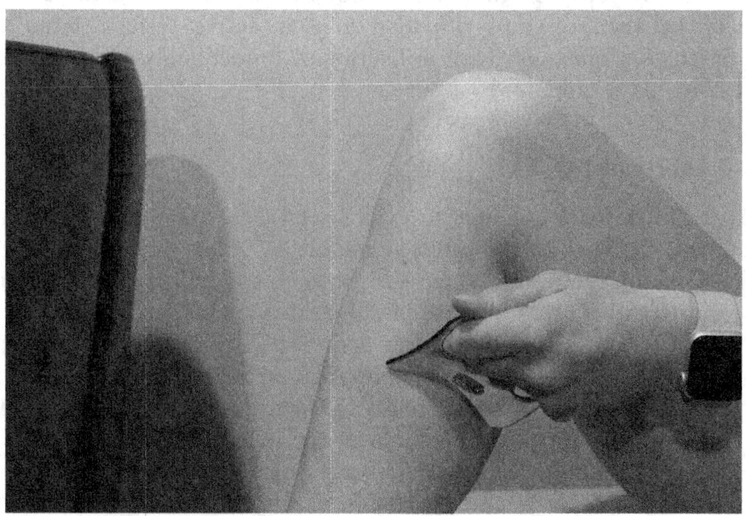

Figure 15.4 Instrument Assisted Soft Tissue Mobilization (IASTM) use on calf

ROM (44). IASTM was reported to provide superior shoulder internal rotation ROM than kinesiotaping (2), and neck ROM compared to routine physical therapy (35).

Training Effects on Range of Motion

IASTM can also have enduring effects after chronic treatments. Repetitive or training-related ROM gains have been reported when testing dorsiflexion after six treatments (10), low back after eight treatments (16), shoulder flexion after once a week treatments for three weeks, shoulder horizontal adduction and internal rotation with five minutes per session, two days a week, over a period of four weeks (20), and one IASTM treatment per week for six weeks significantly improved hip adduction ROM. Again, when comparing ROM benefits to other treatments, IASTM outperformed foam rolling and elastic taping for shoulder ROM when implemented once a week for three weeks (36), and was more beneficial than foam rolling or stretching for hip ROM (once per week for six weeks) (46). IASTM provided similar benefits as a conventional physical therapy programme or kinesiotaping in patients with chronic mechanical low back pain over eight sessions (16), and similar results when compared to a combination of IASTM and post-isometric stretch over four weeks (two days per week) for shoulder adduction (20). Another study reported that five sessions of IASTM with three days of recovery between sessions provided greater active triceps surae ROM than percussive massage (with a Theragun) immediately after training as well as five days later (48).

Global effects of IASTM

We have shown that responses to exercise, fatigue, stretching, or foam rolling are not isolated to the affected muscle or limb, but can have global implications for the performance (i.e., muscle strength, power, and endurance) and ROM of other non-exercised or treated muscles or joints (3, 4, 6, 18, 27, 28). Based on this body of evidence one might expect similar global responses with IASTM. Only one study has investigated non-local IASTM responses. Following four weeks of IASTM on the back (trunk), whether it was applied to the upper or lower part of the superficial back line, there was an increase in supine leg raise (hamstrings) ROM (14). This non-local response may be attributed to cortical responses related to central pain-modulatory systems (14) such as the diffuse noxious inhibitory control system (global release of endorphins and enkephalins in response to pain) (32). Furthermore, there are reports of general body relaxation and decreases in myofascial tone with IASTM (14). In addition, the mechanical force transmission through myofascial chains may transfer strain to neighbouring skeletal muscles and fascia (11).

IASTM Effects on Performance

While stretching research has repeatedly highlighted performance impairments when an acute bout of static stretching is prolonged (i.e., >60-s per muscle group) and not conducted in conjunction with dynamic activities (7, 8, 12, 21), foam rolling typically does not hinder subsequent performance (5, 23, 29). As IASTM is more similar to foam rolling than stretching, what effect would a session of IASTM have on subsequent performance? Similar to many foam rolling studies, there were no immediate deficits apparent in vertical jump height, peak power, and velocity (34), Y-balance test (33) or sprinting speed (47) after an IASTM treatment. In contrast, IASTM alone (once a week for three weeks) (36) or an acute bout in combination with flossing (2) augmented throwing performance. Neck function has been reported to be improved with IASTM and foam rolling (three times in one week) (1) as well as following eight treatments that combined IASTM and exercises for the cervical and thoracic area (38). Similarly, eight IASTM treatments significantly improved the functional disability index (16). Hamstrings power was significantly improved after a single session of IASTM but not with stretching (41). Acute rehabilitation exercises combined with IASTM decreased fatigue and improved isokinetic peak torques (30–180^{0}.s^{-1}) (22). When compared to Theragun percussive massage treatment (five sessions with three days recovery between treatments), IASTM elicited greater improvements in the overhead deep squat immediately after treatment as well as five days later (48). In summary, there are no reports yet of subsequent performance decrements after IASTM, with the possibility of performance enhancements when IASTM is conducted alone or in combination with other treatments such as foam rolling or flossing.

Mechanisms

An increase in pain tolerance is often listed as a major underlying mechanism for the increase in ROM with stretching (3, 7, 8, 12) and foam rolling (24, 25, 26, 27). As you might expect, the discomfort of using a metal tool to repeatedly place stress and strain on your muscles and fascia can affect your pain tolerance. In many instances, there are reports of significant pain reduction or increased pain tolerance with a treatment of IASTM for example, in a single session for 45 seconds at a frequency of 60 beats/minute (chronic neck pain) (15), treatments for low back pain over eight sessions (16), with six treatments over three weeks for myofascial trigger points (17), and four weeks of IASTM reducing headache symptoms and cervical alignment in patients with tension type headaches (43). Modulation of pain tolerance is also seen when IASTM is combined with other modalities such as kinesiotape (16), or foam rolling (three times over one week) (1). Four weeks of IASTM combined with stretching was more effective for neck pain management compared to physical therapy (35). When applied individually, IASTM and foam rolling provided three times over one week were both similarly effective for pain tolerance and sensitivity (1). The concept of myofascial chains or meridians is often explained as a mechanism that transfers mechanical forces, and strains to other regions contributing to decreases in pain and improvements in ROM (11). One study reported that, as IASTM did not alter peak passive torque and muscle stiffness, joint ROM can be improved without altering tissue mechanical properties (19). Thus, similar to stretching and foam rolling, an increase in pain tolerance permits the individual to extend muscles farther with less discomfort in an attempt to increase flexibility.

IASTM may induce inflammation in response to the micro-trauma stress caused by the pressure applied with the metal instruments (39). It has been suggested, but not consistently reported (contradictory results) that the IASTM-induced inflammation would produce accelerated tissue repair, reduce adhesions, induce more rapid collagen formation, and therefore re-establish tissue flexibility (39). In contrast, it has also been proposed that IASTM would decrease inflammation, owing to enhanced lymphatic drainage of toxins (49). However, there is little documentation to support these claims. However, reported increases in blood flow to the treated area (42, 49) would be expected to contribute to an accelerated recovery.

Summary

IASTM provides a similar stimulus to muscle and fascia as foam rolling and thus to be expected provides similar outcomes. IASTM can improve ROM following an acute (minimal volume during a single session of 2–5 minutes) bout as well as demonstrating increased ROM with training (chronic effects occurred with 1–2 sessions per week over 3–6 weeks). This flexibility enhancement is primarily attributed to an increase in pain tolerance as also suggested for stretching and foam rolling. These improvements in ROM are not typically accompanied by performance impairments but typically either no decrements or in some studies performance improvements. IASTM is thus recommended as an adjunct or alternative to other flexibility enhancing techniques.

References

1. Agarwal, S., Bedekar, N., Shyam, A., and Sancheti, P.Comparison between effects of instrument-assisted soft tissue mobilization and manual myofascial release on pain, range of motion and function in myofascial pain syndrome of upper trapezius – A randomized controlled trial. *Hong Kong Physiotherapy Journal*: 1–11, 2024.
2. Angelopoulos, P., Mylonas, K., Tsepis, E., Billis, E., Vaitsis, N., and Fousekis, K. The Effects of Instrument-Assisted Soft Tissue Mobilization, Tissue Flossing, and Kinesiology Taping on Shoulder Functional Capacities in Amateur Athletes. *J Sport Rehabil* 30: 1028–1037, 2021.
3. Behm, D.G., Alizadeh, S., Anvar, S.H., Drury, B., Granacher, U., and Moran, J. Non-local Acute Passive Stretching Effects on Range of Motion in Healthy Adults: A Systematic Review with Meta-analysis. *Sports Med* 51: 945–959, 2021.
4. Behm, D.G., Alizadeh, S., Drury, B., Granacher, U., and Moran J.Non-local acute stretching effects on strength performance in healthy young adults. *Eur J Appl Physiol* 121: 1517–1529, 2021.

5. Behm, D.G., Alizadeh, S., Hadjizadeh Anvar, S., Mahmoud, M.M.I., Ramsay, E., Hanlon, C., and Cheatham, S.Foam Rolling Prescription: A Clinical Commentary. *J Strength Cond Res* 34: 3301–3308, 2020.
6. Behm, D.G., Alizadeh, S., Hadjizedah Anvar, S., Hanlon, C., Ramsay, E., Mahmoud, M.M.I., Whitten, J., Fisher, J.P., Prieske, O., Chaabene, H., Granacher, U., and Steele, J.Non-local Muscle Fatigue Effects on Muscle Strength, Power, and Endurance in Healthy Individuals: A Systematic Review with Meta-analysis. *Sports Med* 51: 1893–1907, 2021.
7. Behm, D.G., Blazevich, A.J., Kay, A.D., and McHugh, M.Acute effects of muscle stretching on physical performance, range of motion, and injury incidence in healthy active individuals: a systematic review. *Appl Physiol Nutr Metab* 41: 1–11, 2016.
8. Behm, D.G. and Chaouachi, A.A review of the acute effects of static and dynamic stretching on performance. *Eur J Appl Physiol* 111: 2633–2651, 2011.
9. Behm, D.G. and Wilke, J.Do Self-Myofascial Release Devices Release Myofascia? Rolling Mechanisms: A Narrative Review. *Sports Med* 49: 1173–1181, 2019.
10. Bush, H.M., Stanek, J.M., Wooldridge, J.D., Stephens, S.L., and Barrack, J.S.Comparison of the Graston Technique(R) With Instrument-Assisted Soft Tissue Mobilization for Increasing Dorsiflexion Range of Motion. *J Sport Rehabil* 30: 587–594, 2020.
11. Carvalhais, V.O., Ocarino, J de M., Araújo, V.L., Souza, T.R., Silva, P.L., and Fonseca, S.T.Myofascial force transmission between the latissimus dorsi and gluteus maximus muscles: an in vivo experiment. *J Biomech* 46: 1003–1007, 2013.
12. Chaabene, H., Behm, D.G., Negra, Y., and Granacher, U.Acute Effects of Static Stretching on Muscle Strength and Power: An Attempt to Clarify Previous Caveats. *Front Physiol* 10: 1468, 2019.
13. Doeringer, J.R., Ramirez, R., and Colas, M.Instrument-Assisted Soft Tissue Mobilization Increased Hamstring Mobility. *J Sport Rehabil* 32: 165–169, 2023.
14. Fousekis, K., Eid, K., Tafa, E., Gkrilias, P., Mylonas, K., Angelopoulos, P., Koumoundourou, D., Billis, V., and Tsepis, E.Can the application of the Ergon((R)) IASTM treatment on remote parts of the superficial back myofascial line be equally effective with the local application for the improvement of the hamstrings' flexibility? A randomized control study. *J Phys Ther Sci* 31: 508–511, 2019.
15. Gercek, H., Unuvar, B.S., Umit Yemisci, O., and Aytar, A.Acute effects of instrument assisted soft tissue mobilization technique on pain and joint position error in individuals with chronic neck pain: a double-blind, randomized controlled trial. *Somatosens Mot Res* 40: 25–32, 2023.
16. Grase, M.O., Elfayez, H.M., Abdellatif, M.M., Genedi, A.F., and Mahmoud, M. A.Effect of instrument assisted soft tissue mobilization versus kinesiotape for chronic mechanical low back pain: a randomized controlled trial. *Physiotherapy Quarterly* 31: 27–33, 2023.
17. Gulick, D.T.Instrument-assisted soft tissue mobilization increases myofascial trigger point pain threshold. *J Bodyw Mov Ther* 22: 341–345, 2018.
18. Halperin, I., Chapman, D.W., and Behm, D.G.Non-local muscle fatigue: effects and possible mechanisms. *Eur J Appl Physiol* 115: 2031–2048, 2015.
19. Ikeda, N., Otsuka, S., Kawanishi, Y., and Kawakami, Y.Effects of Instrument-assisted Soft Tissue Mobilization on Musculoskeletal Properties. *Med Sci Sports Exerc* 51: 2166–2172, 2019.
20. Jusdado-Garcia, M. and Cuesta-Barriuso, R.Soft Tissue Mobilization and Stretching for Shoulder in CrossFitters: A Randomized Pilot Study. *Int J Environ Res Public Health* 18, 2021.

21. Kay, A.D. and Blazevich, A.J.Effect of acute static stretch on maximal muscle performance: a systematic review. *Med Sci Sports Exerc* 44: 154–164, 2012.
22. Kim, J. and Yim, J.Instrument assisted soft tissue mobilization improves physical performance of young male soccer players. *International Journal of Sports Medicine* 39: 936–943, 2018.
23. Konrad, A., Nakamura, M., and Behm, D.G.The Effects of Foam Rolling Training on Performance Parameters: A Systematic Review and Meta-Analysis including Controlled and Randomized Controlled Trials. *Int J Environ Res Public Health* 19, 2022.
24. Konrad, A., Nakamura, M., Bernsteiner, D., and Tilp, M.The Accumulated Effects of Foam Rolling Combined with Stretching on Range of Motion and Physical Performance: A Systematic Review and Meta-Analysis. *J Sports Sci Med* 20: 535–545, 2021.
25. Konrad, A., Nakamura, M., Paternoster, F.K., Tilp, M., and Behm, D.G.A comparison of a single bout of stretching or foam rolling on range of motion in healthy adults. *Eur J Appl Physiol* 122: 1545–1557, 2022.
26. Konrad, A., Nakamura, M., Tilp, M., Donti, O., and Behm, D.G.Foam rolling training effects on range of motion: A Systematic Review and Meta-Analysis. *Sports Med* 52: 2523–2535, 2022.
27. Konrad, A., Nakamura, M., Warneke, K., Donti, O., and Gabriel, A.The contralateral effects of foam rolling on range of motion and muscle performance. *Eur J Appl Physiol* 123: 1167–1178, 2023.
28. Konrad, A., Reiner, M.M., Gabriel, A., Warneke, K., Nakamura, M., and Tilp, M.Remote effects of a 7-week combined stretching and foam rolling training intervention of the plantar foot sole on the function and structure of the triceps surae. *Eur J Appl Physiol* 123: 1645–1653, 2023.
29. Konrad, A., Tilp, M., and Nakamura, M.A Comparison of the Effects of Foam Rolling and Stretching on Physical Performance. A Systematic Review and Meta-Analysis. *Front Physiol* 12: 720531, 2021.
30. Koumantakis, G.A., Roussou, E., Angoules, G.A., Angoules, N.A., Alexandropoulos, T., Mavrokosta, G., Nikolaou, P., Karathanassi, F., and Papadopoulou, M.The immediate effect of IASTM vs. Vibration vs. Light Hand Massage on knee angle repositioning accuracy and hamstrings flexibility: A pilot study. *J Bodyw Mov Ther* 24: 96–104, 2020.
31. Laudner, K., Compton, B.D., McLoda, T.A., and Walters, C.M.Acute effects of instrument assisted soft tissue mobilization for improving posterior shoulder range of motion in collegiate baseball players. *Int J Sports Phys Ther* 9: 1–7, 2014.
32. Le Bars, D., Villanueva, L., Bouhassira, D., and Willer, J.C.Diffuse noxious inhibitory controls (DNIC) in animals and in man. *Patol Fiziol Eksp Ter*: 55–65, 1992.
33. Lehr, M.E., Fink, M.L., Ulrich, E., and Butler, R.J.Comparison of manual therapy techniques on ankle dorsiflexion range of motion and dynamic single leg balance in collegiate athletes. *J Bodyw Mov Ther* 29: 206–214, 2022.
34. MacDonald, N., Baker, R., and Cheatham, S.W.The Effects of Instrument Assisted Soft Tissue Mobilization on Lower Extremity Muscle Performance: A Randomized Controlled Trial. *Int J Sports Phys Ther* 11: 1040–1047, 2016.
35. Mahmood, T., Afzal, W., Ahmad, U., Arif, M.A., and Ahmad, A.Comparative effectiveness of routine physical therapy with and without instrument assisted soft tissue mobilization in patients with neck pain due to upper crossed syndrome. *J Pak Med Assoc* 71: 2304–2308, 2021.
36. Maniatakis, A., Mavraganis, N., Kallistratos, E., Mandalidis, D., Mylonas, K., Angelopoulos, P., Xergia, S., Tsepis, E., and Fousekis, K.The effectiveness of Ergon

Instrument-Assisted Soft Tissue Mobilization, foam rolling, and athletic elastic taping in improving volleyball players' shoulder range of motion and throwing performance: a pilot study on elite athletes. *J Phys Ther Sci* 32: 611–614, 2020.

37. Markovic, G.Acute effects of instrument assisted soft tissue mobilization vs. foam rolling on knee and hip range of motion in soccer players. *J Bodyw Mov Ther* 19: 690–696, 2015.
38. Mylonas, K., Angelopoulos, P., Billis, E., Tsepis, E., and Fousekis, K.Combining targeted instrument-assisted soft tissue mobilization applications and neuromuscular exercises can correct forward head posture and improve the functionality of patients with mechanical neck pain: a randomized control study. *BMC Musculoskelet Disord* 22: 212, 2021.
39. Nazary-Moghadam, S., Yahya-Zadeh, A., Zare, M.A., Ali Mohammadi, M., Marouzi, P., and Zeinalzadeh, A.Comparison of utilizing modified hold-relax, muscle energy technique, and instrument-assisted soft tissue mobilization on hamstring muscle length in healthy athletes: Randomized controlled trial. *J Bodyw Mov Ther* 35: 151–157, 2023.
40. Nielsen, A.Gua Sha: A Clinical Overview. *Chinese Medicine Times* 3: 24–30, 2008.
41. Osailan, A., Jamaan, A., Talha, K., and Alhndi, M.Instrument assisted soft tissue mobilization (IASTM) versus stretching: A comparison in effectiveness on hip active range of motion, muscle torque and power in people with hamstring tightness. *J Bodyw Mov Ther* 27: 200–206, 2021.
42. Portillo-Soto, A., Eberman, L.E., Demchak, T.J., and Peebles, C.Comparison of blood flow changes with soft tissue mobilization and massage therapy. *J Altern Complement Med* 20: 932–936, 2014.
43. Ramadan, S.M., El Gharieb, H.A., Labib, A.M., and Embaby, E.A.Short-term effects of instrument-assisted soft tissue mobilization compared to algometry pressure release in tension-type headache: a randomized placebo-controlled trial. *J Man Manip Ther* 31: 174–183, 2023.
44. Rowlett, C.A., Hanney, W.J., Pabian, P.S., McArthur, J.H., Rothschild, C.E., and Kolber, M.J.Efficacy of instrument-assisted soft tissue mobilization in comparison to gastrocnemius-soleus stretching for dorsiflexion range of motion: A randomized controlled trial. *J Bodyw Mov Ther* 23: 233–240, 2019.
45. Sandrey, M.A., Lancellotti, C., and Hester, C.The Effect of Foam Rolling Versus IASTM on Knee Range of Motion, Fascial Displacement, and Patient Satisfaction. *J Sport Rehabil* 30: 360–367, 2020.
46. Simatou, M., Papandreou, M., Billis, E., Tsekoura, M., Mylonas, K., and Fousekis, K.Effects of the Ergon((R)) instrument-assisted soft tissue mobilization technique (IASTM), foam rolling, and static stretching application to different parts of the myofascial lateral line on hip joint flexibility. *J Phys Ther Sci* 32: 288–291, 2020.
47. Stroiney, D.A., Mokris, R.L., Hanna, G.R., and Ranney, J.D.Examination of Self-Myofascial Release vs. Instrument-Assisted Soft-Tissue Mobilization Techniques on Vertical and Horizontal Power in Recreational Athletes. *J Strength Cond Res* 34: 79–88, 2020.
48. Tyagi, A. and Bhardwaj, A.Effect of IASTM vs. THERAGUN on Triceps Surae Active Range of Motion and Functional Movements in University Level Sprinters. *Indian Journal of Physiotherapy and Occupational Therapy* 16: 1–6, 2022.
49. Vardiman, J.P., Siedlik, J., Herda, T., Hawkins, W., Cooper, M., Graham, Z.A., Deckert, J., and Gallagher, P.Instrument-assisted soft tissue mobilization: effects on the properties of human plantar flexors. *Int J Sports Med* 36: 197–203, 2015.

16 Flossing Effects on Range of Motion and Performance

Floss bands were initially developed by Kelly Starrett (physical therapist) (15). These rubber bands are wrapped around joints or muscles while exercising or stretching (20) (Figure 16.1 and Figure 16.2). Floss bands that can be applied before or after activity are a relatively recent tool purported to ameliorate joint range of motion (ROM) (3, 4, 13, 20), reduce pain (6), prevent sports injuries (25) and improve sports performance (20).

Floss Band Effects on Range of Motion

A number of original research studies have reported ROM improvements with the quadriceps and hamstrings (2, 23, 30), dorsiflexion (5), and plantar flexion ROM (3, 4). Flossing provided greater improvements than dynamic stretching in supine leg raise scores and maximal eccentric knee

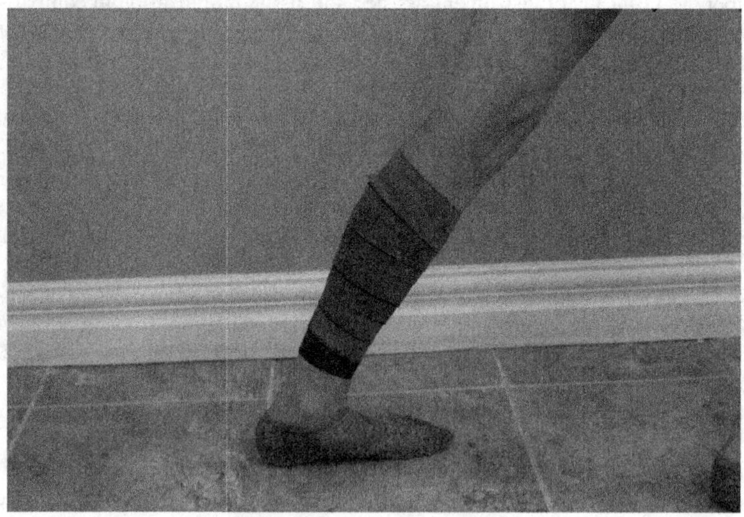

Figure 16.1 Floss band on calf muscles

DOI: 10.4324/9781032709086-18

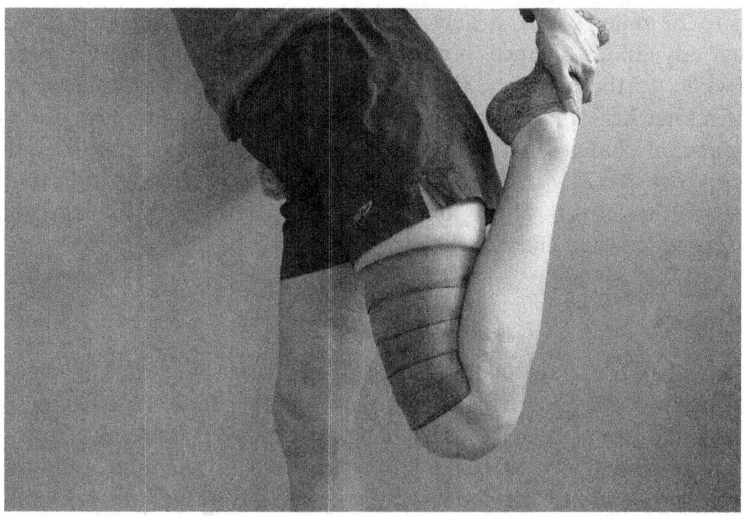

Figure 16.2 Floss band on upper arm

extension contraction force (12). The combination of flossing and active isolated stretching of the hamstrings (three sets of 10 repetitions) tended to be better than stretching without flossing on the sit-and-reach (hamstrings and lower back) test (17). However as with all science, there is always conflicting research with reports of no improvements in plantar flexion or ankle ROM (5, 11), sit and reach scores (9), or the active straight leg raise test (29). So, this is why systematic reviews and meta-analyses were developed. With a meta-analysis we can quantify an overall score based on the literature and the number of participants (more subjects normally suggest a statistically higher-powered study and thus these studies are given more weight in the meta-analysis). Thus, a review of 24 studies by the rising Austrian research star (and former professional soccer/football goalie) Andreas Konrad and colleagues revealed that a single floss band treatment provides small to moderate magnitude increases in ROM. Further sub-group analysis also found weak evidence that flossing during a warm-up may provide greater ROM improvements than stretching, but there was no superior effect when compared to foam rolling (20).

Flossing Effects on Performance

A number of original studies have shown increases in quadriceps isometric force (2, 29), single-leg jump performance (4), countermovement jump and sprint (3, 18, 24), dorsiflexion power (5), agility for one hour after flossing (10), single-leg hop distance and landing stabilization performance in women for up to 20 minutes (30). It may also have no negative effect (30) or actually improve dynamic balance without impairing knee proprioception (2). Meanwhile, to the contrary, there are also reports of floss-induced impairments in knee flexion and extension peak torque, work, and power as well as no significant changes in time to peak torque or flexors–extensors ratio (8) or jump power (23). The Konrad meta-analytical review indicated that flossing can positively affect jumping and strength performance (small to moderate magnitude effects) (20).

Mechanisms

Most floss bands are applied tight enough to restrict blood flow. Hence, some of the mechanisms underlying flossing are reported to be reperfusion, fascial shearing, and blood occlusion (12). The floss-induced ischemia (from occlusion) and subsequent blood flow reperfusion (hyperemia) following its removal may enhance metabolic activity and elevate growth hormone and catecholamine levels (30). Flossing may also increase the pain threshold (e.g., gate control theory, and diffuse noxious inhibitory control release of endorphins and enkephalins) (16, 28), which is also a commonly attributed mechanism for augmenting ROM with stretching (1, 21, 22). The Konrad et al. review (20) suggested that a possible mechanism for floss band-induced increases in ROM or performance is more likely associated with alterations in neuromuscular function (e.g., electromyography, EMG) than muscle mechanical properties (e.g., stiffness). However, contradictory results from other studies indicated that Achilles tendon and gastrocnemius stiffness were reduced for 60 minutes after flossing (18), which may have contributed to improved sprint and jump height, while Pasurka et al. (26) found decreases in ankle capsule and talofibular ligament stiffness as well as reactive hyperemia. An original piece of research from the Konrad group found no floss-induced increases in knee joint ROM but an increase in knee extensors maximum voluntary isometric contraction force (19). As there were no changes in passive resistive torque or EMG, they proposed that there might be an enhancement of growth hormone and norepinephrine after compression release. Another study found that tissue flossing decreased the soleus H-reflex (afferent excitability of the motoneurones) for ten minutes, but this depression did not result in greater ankle joint flexibility or plantar flexor strength (11).

Flossing Effects on Pain

According to the Konrad review (20), there is weak evidence that flossing can relieve pain when treating specific diseases or accelerating recovery after exercise. However, as expected the findings are not unanimous as Gorny and Stoggl (7) reported that flossing did not reduce DOMS symptoms after repetitive leg presses, whereas Prill et al. (27) reported that flossing diminished DOMS (decreased inflammation) by a small magnitude for 48 hours post-exercise. Tissue flossing provided large magnitude improvements with Pain Visual Analog Scale, Short Form McGill Pain Questionnaire II, pain-pressure threshold, hand grip dynamometry and a Likert scale for movement ability in resistance trained individuals with a history of musculoskeletal pain for greater than one month (14).

Summary

Whereas there are always conflicting studies, the overall trend highlights that flossing can increase ROM, improve static (isometric) force, power, and perhaps balance, with some evidence of pain relief. These benefits may stem from a variety of mechanisms related to occlusion-induced reperfusion enhancing metabolic activity and elevating growth hormone and catecholamine levels. In addition, flossing may reduce musculotendinous and ligamentous stiffness and increase pain thresholds.

References

1. Behm, D.G., Blazevich, A.J., Kay, A.D., and McHugh, M.Acute effects of muscle stretching on physical performance, range of motion, and injury incidence in healthy active individuals: a systematic review. *Appl Physiol Nutr Metab 41*: 1–11, 2016.
2. Chang, N.-J., Hung, W.-C., Lee, C.-L., Chang, W.-D., and Wu, B.-H.Effects of a Single Session of Floss Band Intervention on Flexibility of Thigh, Knee Joint Proprioception, Muscle Force Output, and Dynamic Balance in Young Adults. *Applied Sciences* 11: 12052, 2021.
3. Driller, M., Mackay, K., Mills, B., and Tavares, F.Tissue flossing on ankle range of motion, jump and sprint performance: A follow-up study. *Phys Ther Sport* 28: 29–33, 2017.
4. Driller, M.W. and Overmayer, R.G.The effects of tissue flossing on ankle range of motion and jump performance. *Phys Ther Sport* 25: 20–24, 2017.
5. Galis, J. and Cooper, D.J.Application of a Floss Band at Differing Pressure Levels: Effects at the Ankle Joint. *J Strength Cond Res* 36: 2454–2460, 2022.
6. Garcia-Luna, M.A., Cortello-Tormo, J.M., Gonzalez-Martinez, J., and Garcia-Jaen, M.The effects of tissue flossing on perceived knee pain and jump performance: A pilot study. *International Journal of Human Movement and Sport Science* 8: 63–68, 2020.
7. Gorny, V.S. and Stöggl, T.Tissue flossing as a recovery tool for the lower extremity after strength endurance intervals. *Sportverl Sportschad* 32: 55–60, 2018.
8. Hadamus, A., Jankowski, T., Wiaderna, K., Bugalska, A., Marszalek, W., Blazkiewicz, M., and Bialoszewski, D.Effectiveness of Warm-Up Exercises with Tissue Flossing in Increasing Muscle Strength. *J Clin Med* 11, 2022.
9. Hadamus, A., Kowalska, M., Kedra, M., Wiaderna, K., and Bialoszewski, D.Effect of hamstring tissue flossing during warm-up on sit and reach performance. *J Sports Med Phys Fitness* 62: 51–55, 2022.
10. Huang, Y.S., Lee, C.L., Chang, W.D., and Chang, N.J.Comparing the effectiveness of tissue flossing applied to ankle joint versus calf muscle on exercise performance in female adults: An observational, randomized crossover trial. *J Bodyw Mov Ther* 36: 171–177, 2023.

11. Kalc, M., Mikl, S., Zoks, F., Vogrin, M., and Stoggl, T.Effects of Different Tissue Flossing Applications on Range of Motion, Maximum Voluntary Contraction, and H-Reflex in Young Martial Arts Fighters. *Front Physiol* 12: 752641, 2021.
12. Kaneda, H., Takahira, N., Tsuda, K., Tozaki, K., Kudo, S., Takahashi, Y., Sasaki, S., and Kenmoku, T.Effects of Tissue Flossing and Dynamic Stretching on Hamstring Muscles Function. *J Sports Sci Med* 19: 681–689, 2020.
13. Kaneda, H., Naonobu, T., Kouji, T., Kiyoshi, T., Kenta, S., Sho, K., Yoshiki, T., Shuichi, S., Kensuke, F., and Tomonori, K.The effects of tissue flossing and static stretching on gastrocnemius exertion and flexibility. *Isokinetics and Exercise Science* 28: 205–213, 2020.
14. Kelly, C.F., Oliveri, Z., Saladino, J., Senatore, J., Kamat, A., Zarour, J., and Douris, P.C.The Acute Effect of Tissue Flossing on Pain, Function, and Perception of Movement: A Pilot Study. *Int J Exerc Sci* 16: 855–865, 2023.
15. Kiefer, B., Lemarr, K.E., Enriquez, C., and Tivener, K.A pilot study: Psychological effects of the voodoo floss band on glenohumeral flexibility. *International Journal of Athletic Therapy and Training* 22: 1–16, 2021.
16. Kielur, D.S. and Powden, C.J.Changes of ankle dorsiflexion using compression tissue flossing: A systematic review and meta-analysis. *Journal of Sport Rehabilitation* 30: 306–314, 2021.
17. Kitsuksan, T. and Earde, P.The immediate effects of tissue flossing during active isolated stretching on hamstring flexibility in young healthy individuals. *Physiotherapy Quarterly* 30: 61–67, 2021.
18. Klich, S., Smoter, M., Michalik, K., Bogdanski, B., Valera Calero, J.A., Manuel Clemente, F., Makar, P., and Mroczek, D.Foam rolling and tissue flossing of the triceps surae muscle: an acute effect on Achilles tendon stiffness, jump height and sprint performance – a randomized controlled trial. *Res Sports Med*: 1–14, 2022.
19. Konrad, A., Bernsteiner, D., Budini, F., Reiner, M.M., Glashuttner, C., Berger, C., and Tilp, M.Tissue flossing of the thigh increases isometric strength acutely but has no effects on flexibility or jump height. *Eur J Sport Sci* 21: 1648–1658, 2021.
20. Konrad, A., Mocnik, R., and Nakamura, M.Effects of Tissue Flossing on the Healthy and Impaired Musculoskeletal System: A Scoping Review. *Front Physiol* 12: 666129, 2021.
21. Magnusson, S.P. and Renstrom, P.The European College of Sports Sciences Position statement: The role of stretching exercises in sports. *European Journal of Sport Science* 6: 87–91, 2006.
22. Magnusson, S.P., Simonsen, E.B., Aagaard, P., Sorensen, H., and Kjaer, M.A mechanism for altered flexibility in human skeletal muscle. *Journal of Physiology* 497 (Pt 1): 291–298, 1996.
23. Maust, Z., Bradney, D., Collins, S.M., Wesley, C., and Bowman, T.G.The Effects of Soft Tissue Flossing on Hamstring Range of Motion and Lower Extremity Power. *Int J Sports Phys Ther 16*: 689–694, 2021.
24. Mills, B., Mayo, B., Tavares, F., and Driller, M.The Effect of Tissue Flossing on Ankle Range of Motion, Jump, and Sprint Performance in Elite Rugby Union Athletes. *J Sport Rehabil* 29: 282–286, 2020.
25. Mohamed, E.E., Useh, U., and Mtshali, B.F. Q-angle, pelvic width, and intercondylar notch width as predictors of knee injuries in women soccer players in South Africa. *African Health Science* 12: 174–180, 2012.
26. Pasurka, M., Lutter, C., Hoppe, M.W., Heiss, R., Gaulrapp, H., Ernstberger, A., Engelhardt, M., Grim, C., Forst, R., and Hotfiel, T.Ankle flossing alters

periarticular stiffness and arterial blood flow in asymptomatic athletes. *J Sports Med Phys Fitness* 60: 1453–1461, 2020.

27. Prill, R., Schulz, R., and Michel, S.Tissue flossing: a new short-term compression therapy for reducing exercise-induced delayed-onset muscle soreness. A randomized, controlled and double-blind pilot crossover trial. *J Sports Med Phys Fitness* 59: 861–867, 2019.
28. Stevenson, P.J., Stevenson, R.K., and Duarte, K.W.Acute effects of the voodoo flossing band on ankle range of motion. *Journal of Medicine, and Biomedical Applied Science*7: 244–253, 2019.
29. Vogrin, M., Kalc, M., and Ličen, T.Acute Effects of Tissue Flossing Around the Upper Thigh on Neuromuscular Performance: A Study Using Different Degrees of Wrapping Pressure. *Journal of Sport Rehabilitation* 30: 601–608, 2021.
30. Wu, S.Y., Tsai, Y.H., Wang, Y.T., Chang, W.D., Lee, C.L., Kuo, C.A., and Chang, N.J.Acute Effects of Tissue Flossing Coupled with Functional Movements on Knee Range of Motion, Static Balance, in Single-Leg Hop Distance, and Landing Stabilization Performance in Female College Students. *Int J Environ Res Public Health* 19, 2022.

17 Stretching Exercise Illustration

This chapter provides photographs with explanations of a variety of stretching exercises that can be employed within a standard athletic warm-up or within a dedicated flexibility training sessions. A standardized warm-up is presented below which outlines in general the allocation of flexibility, mobility and activation activities.

Table 17.1 Standardized Warm-up Components Example

Submaximal Aerobic Warm-up	*Static Stretching*	*Mobility and Activation*	*Dynamic Stretching and Sport Specific Movements*
Examples: Running, jogging, bike (arms and legs), cycle ergometer, skipping. Elliptical, attempt to increase core temperature by 1–2°C as evidenced by a sweating response	1–2 repetitions of 15–30 seconds per muscle group (≤60 seconds per muscle group) 3–5 stretches Target restricted groups or relevant groups for training session	Choose two or three exercises to address weaknesses and 2–3 exercises for specific movement patterns 5–8 repetitions per exercise Address common weaknesses or deficiencies (e.g. gluteal bridge, band hip flexor, banded lateral/forward walks, band pull-aparts) Prepare movement patterns for workout (e.g. squat, hinge)	5–8 exercises, Example: 10-minute sprint out and back Elevate heart rate, prepare for workout, work on movement patterns
5–10 minutes	3–5 minutes maximum	3–6 minutes maximum	3 minutes maximum

DOI: 10.4324/9781032709086-19

Static Stretching Exercise Section

Lower Body Stretching Photos: Quadriceps and Hip Flexors

Figure 17.1 Kneeling Hip Flexor with Foot Elevated Stretch

Figure 17.2 Forward Lunge Stretch with Hand Planted

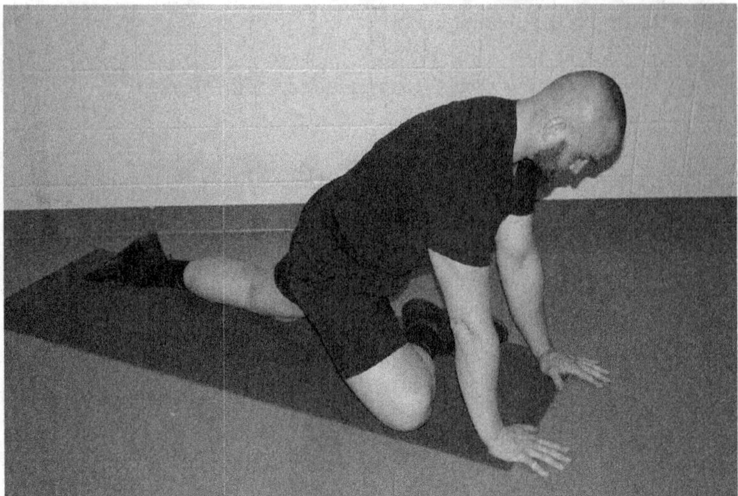

Figure 17.3 Pigeon Stretch

Figure 17.4 Lunge with a rotation

Lower Body Stretching Photos: Hamstrings

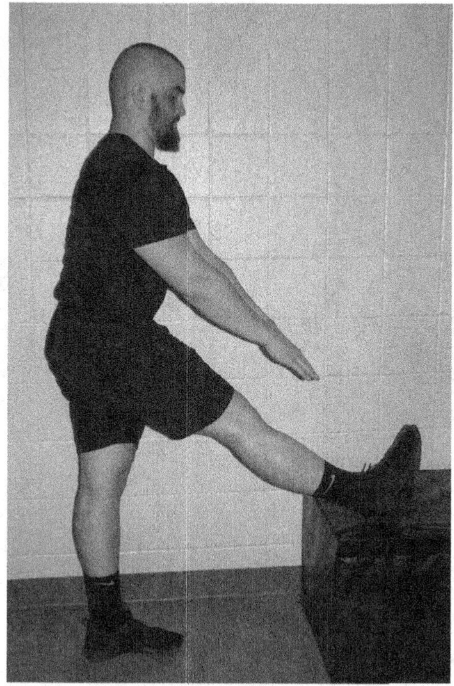

Figure 17.5 Single leg elevated hamstrings stretch

Figure 17.6 Supine single leg hamstring stretch

Figure 17.7 Band Assisted Hamstring Stretch

Lower Body Stretching Photos: Gluteals

Figure 17.8 Reverse Pigeon Stretch

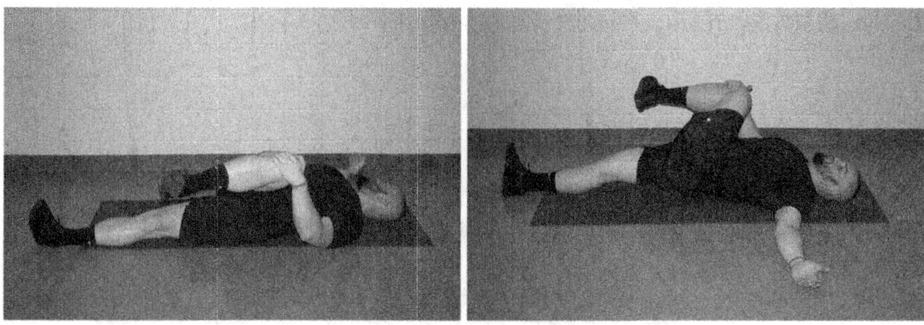

Figure 17.9 Supine single leg rotation

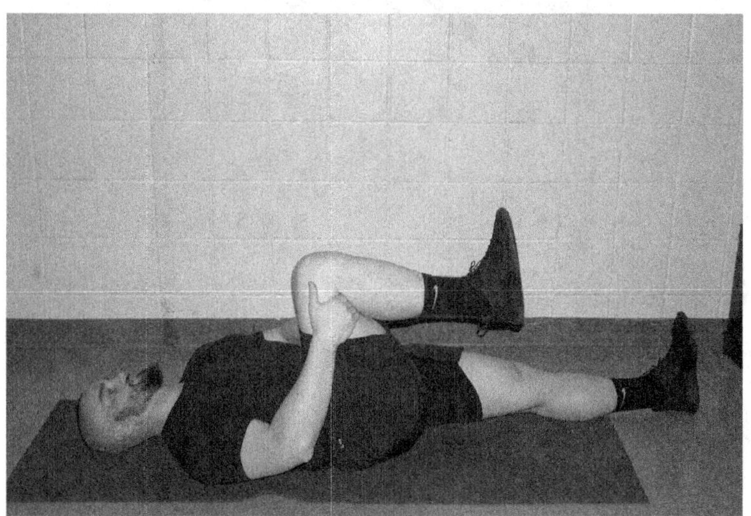

Figure 17.10 Lying (supine) hip flexion

Lower Body Stretching Photos: Groin Adductors

Figure 17.11 Adductor Stretch

Lower Body Stretching Photos: Calves (Plantar flexors)

Figure 17.12 Push up Position Calf Stretch

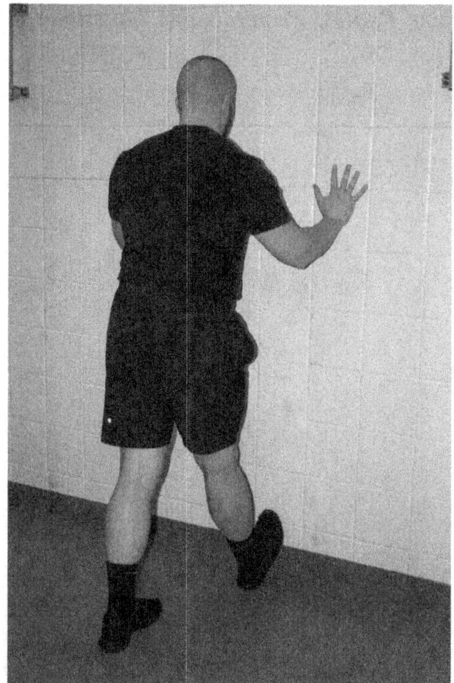

Figure 17.13 Standing Calf Stretch Against Wall

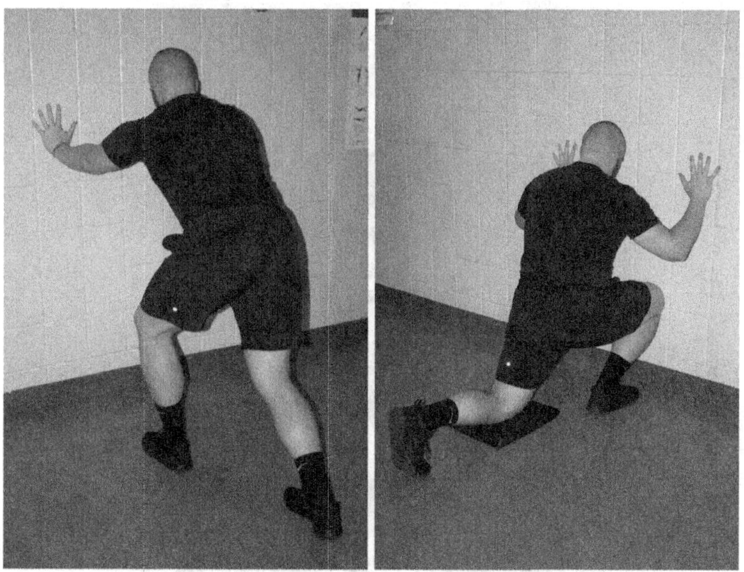

Figure 17.14 Standing Soleus Stretch

Upper Body Stretching Photos: Pectorals and Latissimus Dorsi

Figure 17.15 Pectoralis Wall Stretch with Rotation

Figure 17.16 Band Assisted Pectoralis Stretch

Figure 17.17 Banded Latissimus Dorsi Stretch

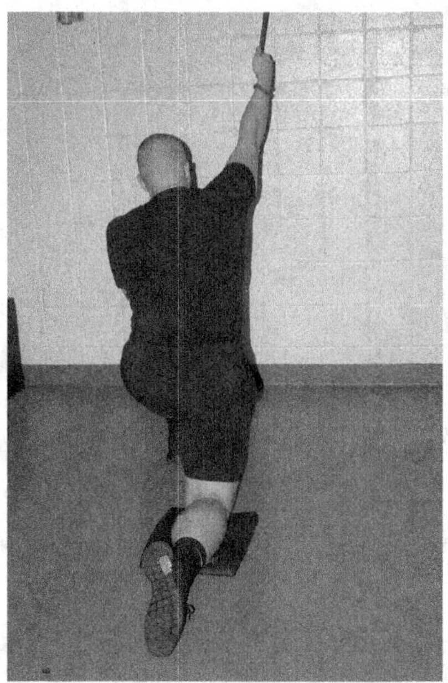

Figure 17.18 Kneeling lunge position latissimus dorsi overhead stretch with band

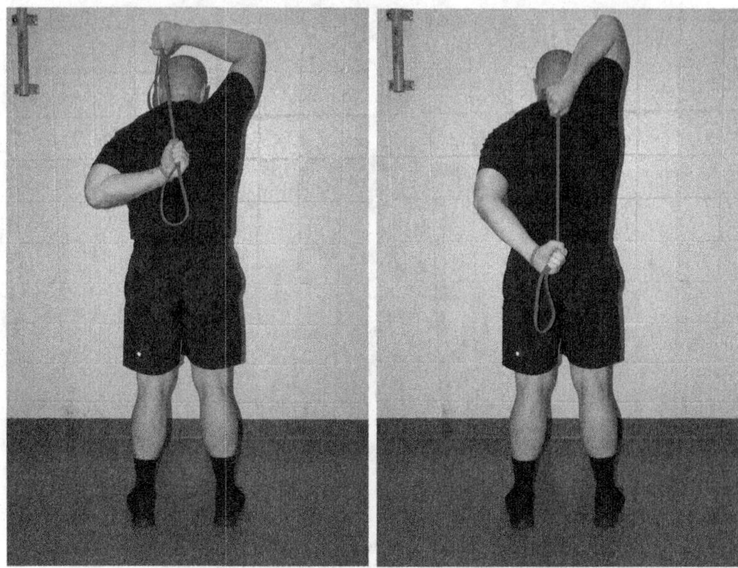

Figure 17.19 a,b Triceps brachii and shoulder internal rotation band stretch

Figure 17.20 Shoulder External rotation with band

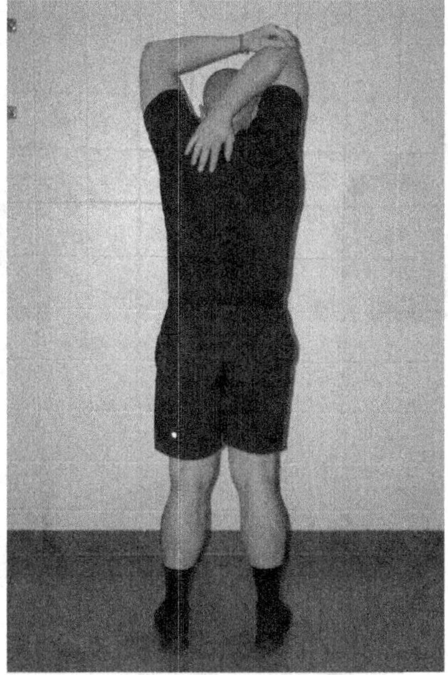

Figure 17.21 Overhead triceps brachii stretch

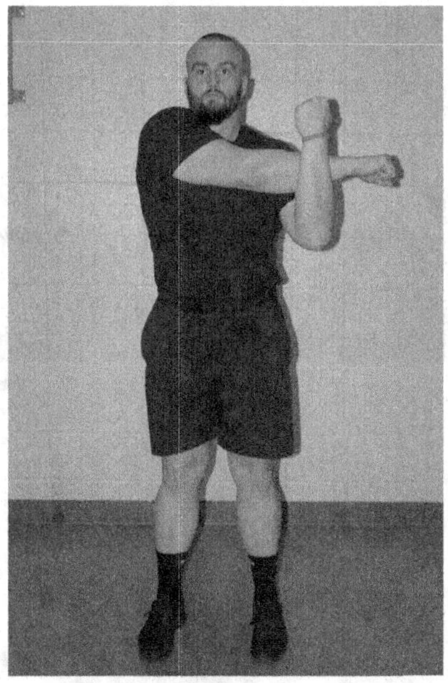

Figure 17.22 Deltoid cross-over stretch

Figure 17.23 a,b,c Neck series

Upper Body Stretching Photos: Neck

Figure 17.24 a,b,c Neck strengthening series

Figure 17.25 a,b Hip Mobility Hinge and Overhead Squat Sequence

Mobility Exercise Section

Lower Body Mobility Photos

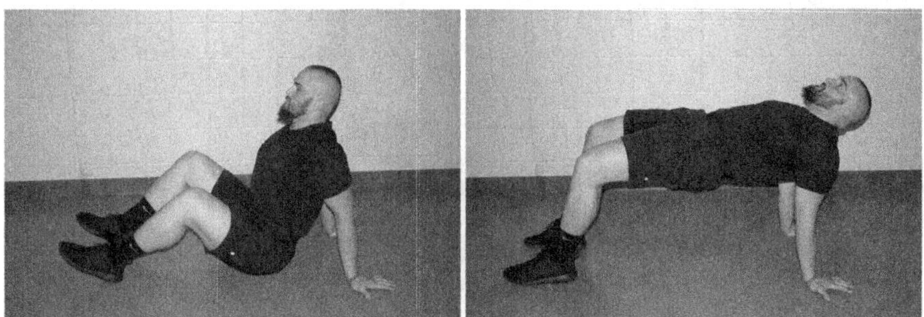

Figure 17.26 a,b Dynamic hip bridge with arms extended

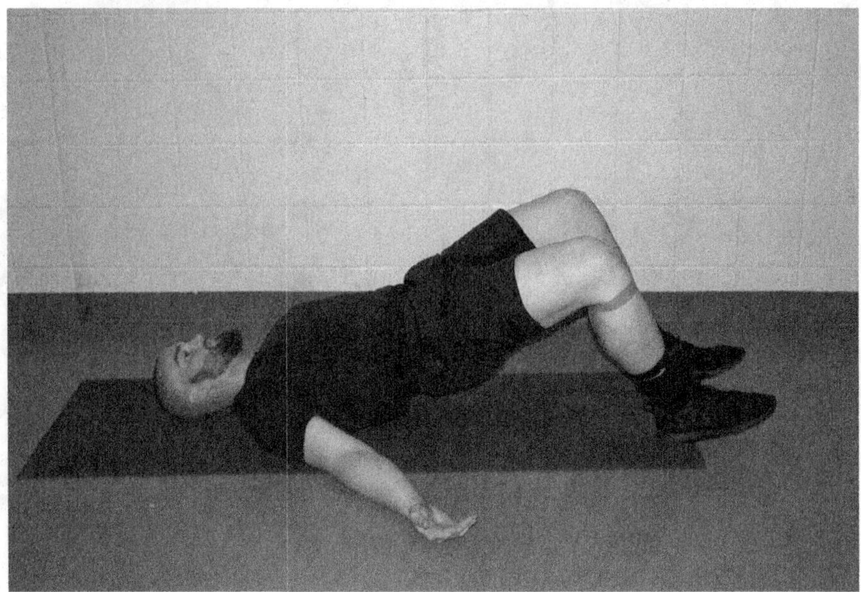

Figure 17.27 Dynamic Hip Bridge with Band

Figure 17.28 Dynamic Alternating Hip Bridge with a Knee to Chest Hold

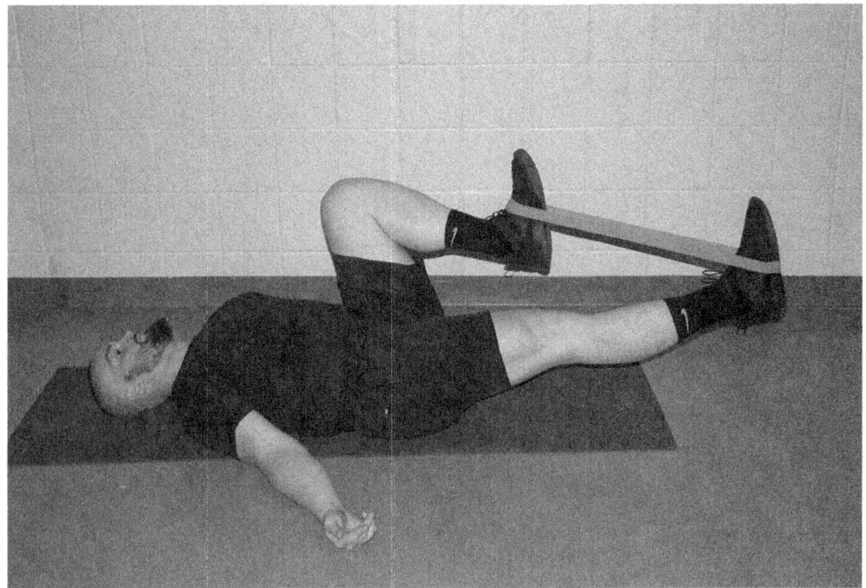

Figure 17.29 Dynamic dead bug with band

Figure 17.30 Dynamic standing alternating knee to chest

Figure 17.31 Dynamic lateral lunge with pause

Figure 17.32 Dynamic ankle to opposite hip with pause

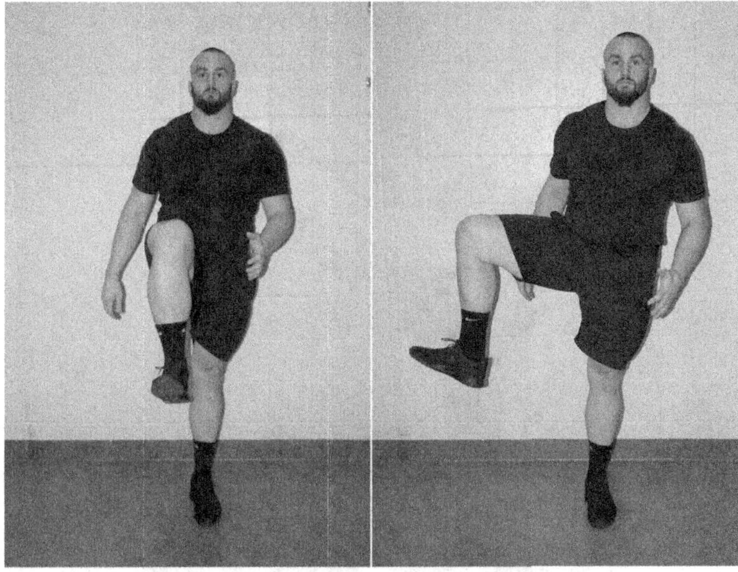

Figure 17.33 a,b Dynamic external rotation of hip

Figure 17.34 a,b Dynamic single leg hinge with arm reach series

Figure 17.35 Dynamic leg kicks with hand touch

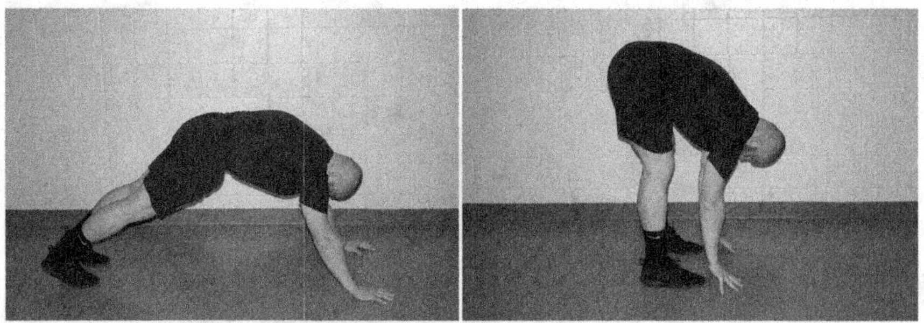

Figure 17.36 a,b Dynamic inch worm with pause

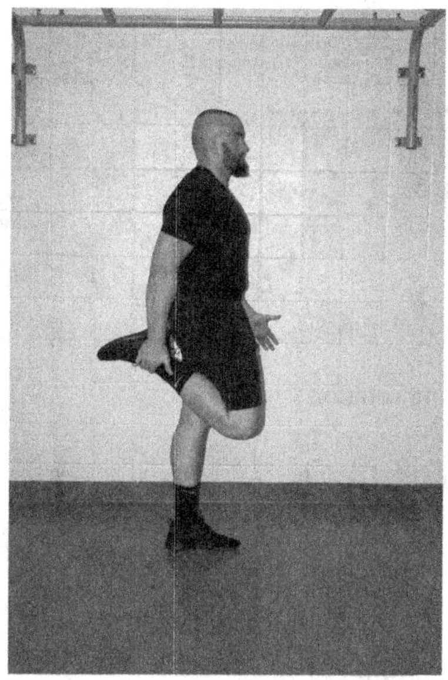

Figure 17.37 Dynamic quadriceps squeeze with a pause

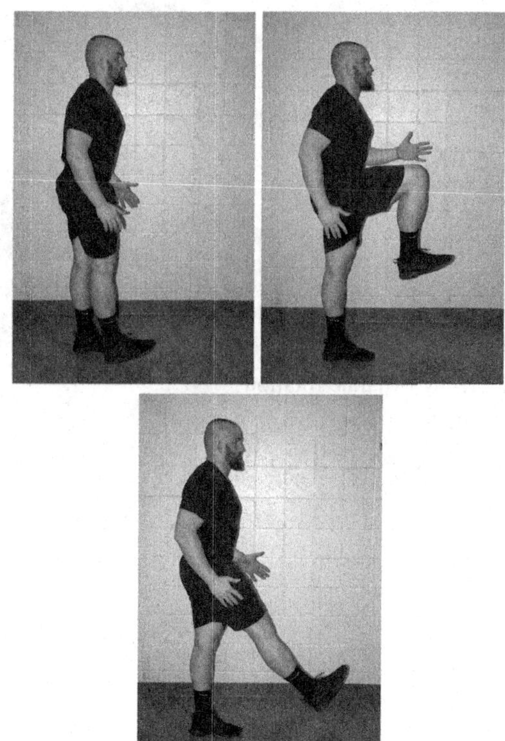

Figure 17.38 a,b,c Sprinting mechanics

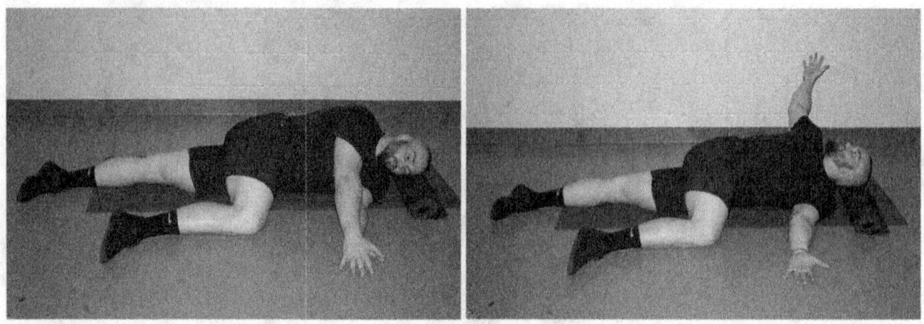

Figure 17.39 a,b Side lying rotation

Upper Body Mobility Photos

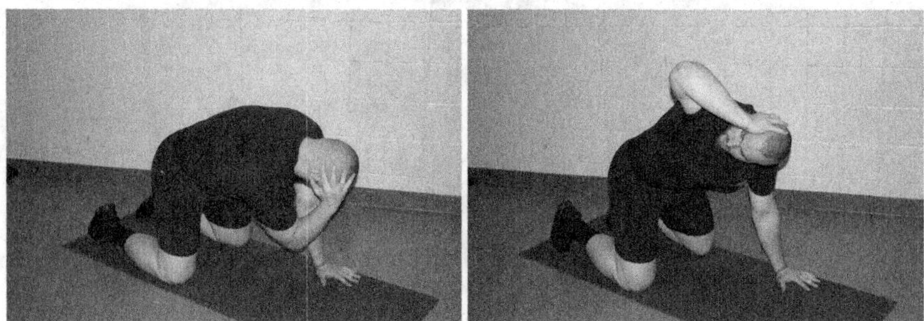

Figure 17.40 a,b Kneeling dynamic rotation

Figure 17.41 a,b,c Dynamic band shoulder pullovers

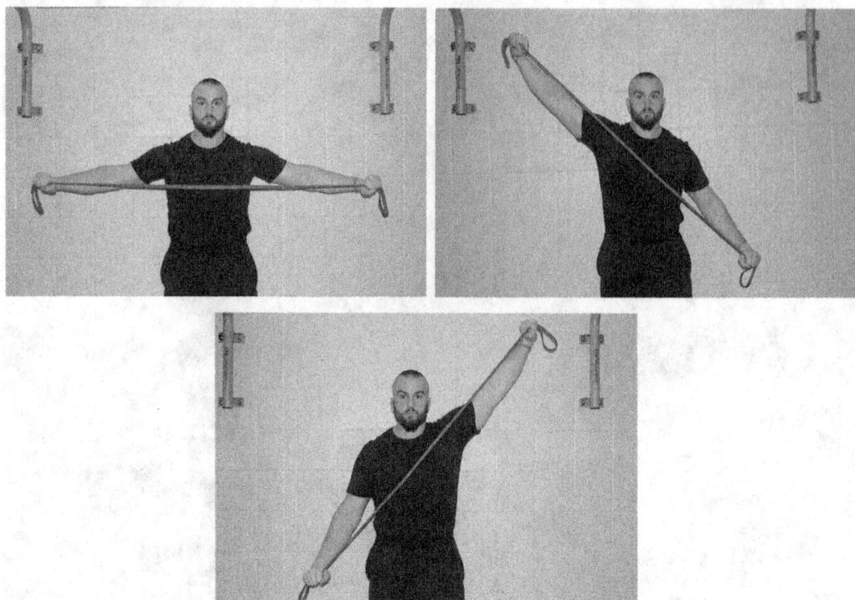

Figure 17.42 a,b,c Band pull apart series

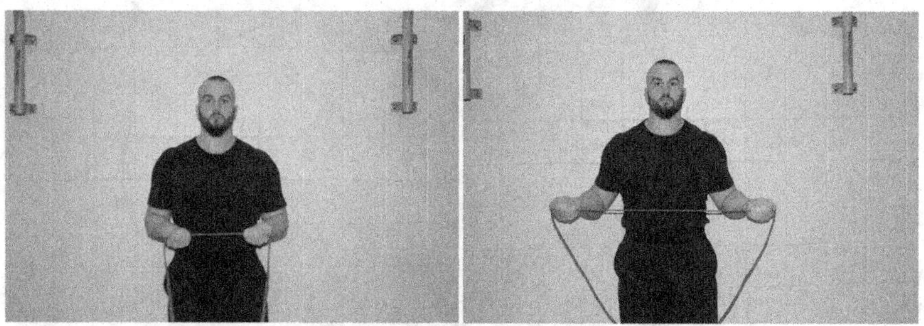

Figure 17.43 a,b Band shoulder external rotation

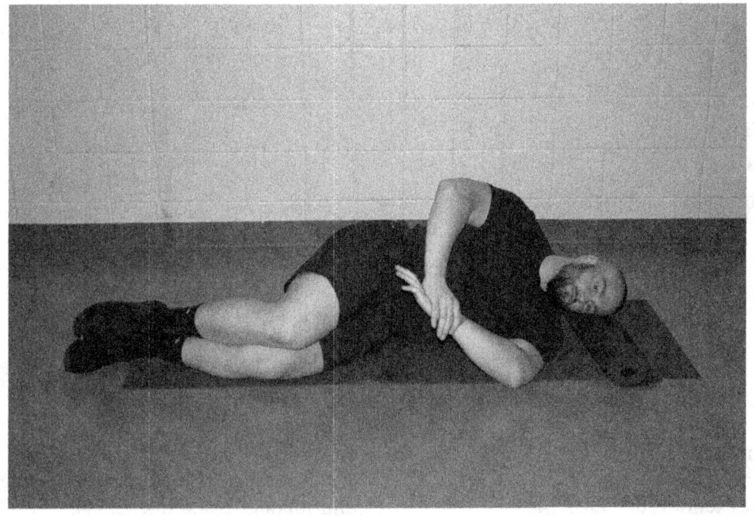

Figure 17.44 Shoulder External rotation sleep stretch

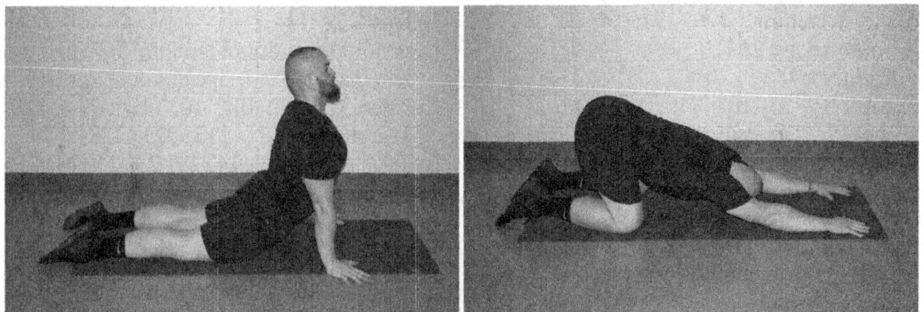

Figure 17.45 a,b Dynamic prone cobra to child's pose

Index

Note: locators in *italics* and **bold** refer to *figures* and **tables** respectively.

Printed in the United States
by Baker & Taylor Publisher Services